SOIL PHYSICS

McGRAW-HILL PUBLICATIONS IN THE AGRICULTURAL SCIENCES

Adriance and Brison · Propagation of Horticultural Plants
Ahlgren · Forage Crops
Anderson · Diseases of Fruit Crops
Brown and Ware · Cotton
Carroll, Krider, and Andrews · Swine Production
Christopher · Introductory Horticulture
Crafts and Robbins · Weed Control
Cruess · Commercial Fruit and Vegetable Products
Dickson · Diseases of Field Crops
Eckles, Combs, and Macy · Milk and Milk Products
Elliott · Plant Breeding and Cytogenetics
Fernald and Shepard · Applied Entomology
Gardner, Bradford, and Hooker · The Fundamentals of Fruit Production
Gustafson · Conservation of the Soil
Gustafson · Soils and Soil Management
Hayes, Immer, and Smith · Methods of Plant Breeding
Herrington · Milk and Milk Processing
Jenny · Factors of Soil Formation
Jull · Poultry Husbandry
Kohnke · Soil Physics
Kohnke and Bertrand · Soil Conservation
Laurie and Ries · Floriculture
Leach · Insect Transmission of Plant Diseases
Maynard and Loosli · Animal Nutrition
Metcalf, Flint, and Metcalf · Destructive and Useful Insects
Nevens · Principles of Milk Production
Paterson · Statistical Technique in Agricultural Research
Peters and Grummer · Livestock Production
Rather and Harrison · Field Crops
Rice, Andrews, Warwick, and Legates · Breeding and Improvement of Farm Animals
Roadhouse and Henderson · The Market-milk Industry
Steinhaus · Principles of Insect Pathology
Thompson · Soils and Soil Fertility
Thompson and Kelly · Vegetable Crops
Thorne · Principles of Nematology
Tracy, Armerding, and Hannah · Dairy Plant Management
Walker · Diseases of Vegetable Crops
Walker · Plant Pathology
Wilson · Grain Crops
Wolfe and Kipps · Production of Field Crops

SOIL PHYSICS

HELMUT KOHNKE
Soil Scientist
Purdue University

McGraw-Hill Book Company
New York St. Louis
San Francisco Toronto
London Sydney

SOIL PHYSICS

Library of Congress Catalog Card Number 68-21846

35299

1 2 3 4 5 6 7 8 9 0 M A M M 7 5 4 3 2 1 0 6 9 8

PREFACE

For some time there has been a need for a comprehensive yet concise textbook on soil physics. Improvement of fertilization, crop varieties, and technology has increased farm yields so much that frequently soil physical conditions have become the limiting factors of crop production. The exigencies of soil erosion control, irrigation, management of greenhouse soils, and the selection of proper land for residential purposes also require a clear understanding of soil physics.

Recognizing this situation, colleges are putting increased emphasis on the teaching of soil physics. This book has been prepared to serve as a text for such courses as well as for self-study. The subject matter is organized to give first a picture of the physical-chemical properties of water and of the interactions between soil and water. Next the solid components of the soil are discussed. Building upon this background, chapters on soil air, soil temperature, and soil color are presented. The final chapter is devoted to the application of soil physics in the management of the land. Basic knowledge of physics, chemistry, and soil science is assumed.

It is a pleasure to express my gratitude for the assistance received through stimulating discussions with my colleagues. These include Professors S. A. Barber, P. F. Low, P. G. Moe, J. E. Newman, A. J. Ohlrogge, D. Swartzendruber, J. L. White, and D. Wiersma of Purdue University, and Professor C. I. Rich of Virginia Polytechnic Institute.

Helmut Kohnke

CONTENTS

1 INTRODUCTION

WHAT IS SOIL PHYSICS?

Soils are highly complicated systems and it is necessary to define the approach which is used to study them. One of the fundamental subjects of soil science is soil physics, the study of mechanics, heat, and optics as they relate to soil. However, many physical-chemical properties of the soil are also usually included under soil physics, such as thermodynamics, colloidal behavior, and liquids. No text on soil physics would be complete that did not present a discussion of the effects of soil physical conditions on plant growth and of the management of soil physical conditions.

THE PLACE OF SOIL PHYSICS IN SOIL SCIENCE

The purpose of agronomic, horticultural, and sylvicultural soil science is the growing of plants. Engineering soil science obviously has other interests. Whether we still believe with Justus von Liebig the law of the minimum, or whether we have accepted Mitscherlich's law of the plant-growth factors, or if we have only a casual acquaintance with the subject, we realize that the cooperation of many heterogeneous factors is necessary to produce a plant.

In order to understand all these factors the study of soil science has been divided into various fields: soil physics, soil chemistry, soil mineralogy, soil microbiology, soil fertility, soil genesis, soil morphology, classification and survey, soil technology, and soil conservation. There is no definite boundary between these fields of soil science, nor can any one be assumed to be more important than the others. No individual branch of soil science is of value unless it is related to the other branches.

In the last analysis all of these are studied with the idea of maintaining or increasing the productivity of the land. A few examples will illustrate the relationship of soil physics to productivity. A soil rich in all necessary chemicals is a desert, if water is absent. A moist and fertile bottom land becomes a worthless swamp if inundation removes the oxygen supply that the roots need. The resources of the subsoil remain inaccessible to the crops if a dense plow sole inhibits root penetration.

THE PURPOSE OF THE STUDY OF SOIL PHYSICS

The purpose of studying soil physics is to get acquainted with the important phases of this subject and to acquire tools with which to tackle the problems of crop production. These tools are both abstract and concrete. They include a realization of the laws governing the physical behavior of soil as well as the ability to use the methods necessary to determine the physical properties of the soil, and the judgment how to modify physical conditions of soil in the field.

Summarizing, it can be stated:

The purpose of the study of soil physics is to

Provide information on the fundamentals of soil physics,

Become acquainted with the effects of soil physics on plant growth,

Develop a working knowledge of some of the methods and instrumentation used in soil-physics research,

Learn to use soil physics in evaluating soils,

Learn to evaluate the influences of the environmental factors on soil properties,

Show how physical conditions of soils can be influenced to serve mankind, and

Help to develop a philosophy of soil science.

HISTORIC BACKGROUND OF SOIL PHYSICS

Soil physics has received less attention than soil chemistry and soil fertility. What was the reason for this lack of popularity? Up to about 100 years ago commercial fertilizers were not yet widely known. A farmer who wanted to increase his yields had to do it by good tillage, drainage, irrigation, rotation, manuring, and liming. With the introduction of guano, superphosphate, and later saltpeter from Chile he could—on the impoverished fields—obtain much more spectacular yield increases than by the tried and true methods of physical improvement of the soil. It is no wonder that he neglected the art of tillage more and more and that when his son went to college, he wanted to know more about these miracle salts that could be sprinkled on the ground than about factors that would affect the tilth and the tillage of the land, which he could learn directly from his father.

Another reason for the predominance of chemistry over physics in soil science from 1840 to 1920 is that many of the leaders in the field were chemists and geologists by training. An outstanding influence in directing soil science toward the chemical viewpoint was exerted by Justus von Liebig. His students and his writing, especially his book "Organic Chemistry in Its Application to Agriculture and Physiology," which was published in 1840, carried the message of soil chemistry throughout the world. There was no equivalent counterpart in soil physics in that period.

Compared to the chemical and biologic phenomena, the physical phenomena of soils, such as texture, moisture control, and temperature, appear deceptively simple to the casual observer and may not have been considered as a challenge by some investigators. On the other hand, research into soil physics requires such strenuous field work that this may have deterred scientifically trained men. It is only relatively recently that soil physics has gotten into the foreground again. Three factors are mostly responsible for this:

1. The fact that in many cases the best-known application of commercial fertilizers does not suffice to ensure profitable yields.
2. The recognition of the soil profile as a natural body and an entity.
3. The washing away of much soil by erosion that became more and more apparent and has received much publicity.

As with many other fields, it is difficult to say when soil physics started as an individual science. A good deal was known about the physical conditions of soils in ancient Greece and Rome. No "father" of soil physics stands out. One of the first men in modern times to study soil-water relationships was De la Hire, who in 1688 published a treatise on percolation of water through lysimeters. He established the fact that rain is the origin of the water in the springs. De la Hire was mathematician and meterologist at the court of Louis XIV at Versailles.

Schübler reported the first systematic study of a wide variety of soil physical properties in 1833. He was particularly interested in those properties that might affect crop yields. In 1864 Schumacher introduced the concept of capillary and noncapillary porosity of soils; he recognized the effect of shading of plant leaves on soil and the significance of protection offered by foliage against beating action of raindrops.

Wollny was one of the outstanding pioneers in the field of soil physics. During the period from 1879 to 1898 he published the results of a variety of soil physical research in the *Forschungen auf dem Gebiete der Agriculturphysik*, a periodical of which he was the editor. His investigations in the field of hydrology in relation to soil physics are classical. Although today of historic value only, it is interesting to find in these publications ideas and conclusions that were considered a discovery in a period about 50 years later.

Some of the more important exponents of soil physics in America during the second half of the nineteenth century were Hilgard in California, Johnson in Connecticut, King in Wisconsin, and Slichter in the U.S. Geological Survey. These men were general soil scientists stressing soil physics in part of their work.

During this century soil physics has come into its own. Nevertheless, up to about 1930 the majority of soil physical analyses were done on sieved soil, evidently because the importance of natural structure had not yet been recognized.

Some of the scientists, active in the first third of the twentieth century, who have contributed much to the advancement of soil physics are listed here:

Briggs moisture equivalent,
Buckingham capillary potential,
Patten transference of heat in soils,
Mitscherlich hygroscopicity, heat of wetting,
Lebedev condensation and adsorption of water vapor by the soil,
Bouyoucos mechanical analysis with hydrometer, moisture measurement by electrical resistance,
Sekera, Donat, Schofield soil-moisture potential,
Bradfield nature of colloidal clay,
Kubiena microstructure of soil.

More recent workers in soil physics are too numerous to mention.

SOIL–PHYSICS LITERATURE

Soil physics is a rapidly developing science. In order to become acquainted with the developments in the different branches of soil physics, the reader should consult the recent volumes of the specific technical literature. The following list of publications is recommended as a source of additional information.

Advances in Agronomy,
Agricultural Meteorology,
Agronomy Journal,
Annales Agronomiques,
Australian Journal of Soil Research,
Canadian Journal of Soil Science,
Journal of Agricultural Science,
Journal of Hydrology,
Journal of Soil and Water Conservation,
Journal of Soil Science,
Mitteilungen der Deutschen Bodenkundlichen Gesellschaft,
Soil Science,
Soil Science Society of America Proceedings,
Soviet Soil Science (Pochvovedeniye),
Transactions of the ASAE (American Society of Agricultural Engineers),
Transactions of the International Congresses of Soil Science,
Water Resources Research,
Zeitschrift für Pflanzenernährung und Bodenkunde.

2 | SOIL WATER

One of the most important ingredients of the soil is the moisture that fills part of the pores between the solid particles. It is also one of its most dynamic properties. Water affects intensely many physical and chemical reactions of the soil as well as plant growth. Therefore, a knowledge of the behavior of soil water is fundamental for the understanding of most soil physical phenomena.

PROPERTIES OF WATER

THE IMPORTANCE OF WATER

Along with air, water may be considered as the most basic resource. No agriculture and no industry, in fact no life is possible without water. The amounts of water required to produce such everyday articles as milk, steel, wood, and plastic are staggering. It is obvious that water is needed in the production of agricultural goods, but it is equally true that no industrial item can be manufactured without water.

Of particular interest to us is water as part of the soil. It serves in the nutrition of plants and microbes and in the development of the soil. It is very important that water is of good quality for human consumption, for agriculture, and for industry. Soil plays a profound role in providing pure water, by filtering out solids and adsorbing undesirable chemicals. On the other hand, it may be responsible for making surface runoff water muddy or full of objectionable salts.

THE OCCURRENCE OF WATER

Water is ubiquitous. It is an integral part of every living being and is an essential component of fertile soil. It occurs from great depths in the earth's crust to the stratosphere. Over 99 percent of all of the earth's water occurs in the troposphere. This reaches up to an average height of 15 km. Above this to about 30 km is the stratosphere. This contains very little water, always in the form of ice, visible as cirrus clouds. Water molecules cannot exist in the chemosphere (30 to 80 km). Short-wave radiation is so intense at this height that molecules are broken up into atoms. Beyond the chemosphere is the ionosphere, where atoms lose their electrons and become ionized.

The average depth to which water occurs in the ground is about 3 km, in some locations as deep as 8 km. The ocean floor descends deeper than that. The density of the rocks increases with depth, decreasing the pore space available for water, until finally there is none and the rock becomes viscous. The high temperatures of the lower strata would also prevent the presence of water.

THE PHYSICOCHEMICAL NATURE OF WATER, ITS STRUCTURE, AND ITS THREE STATES

The space occupied by an individual water molecule is largely determined by that of the oxygen ion. The two hydrogen ions take up practically no space. Therefore the diameter of a water molecule is approximately 2.64 Å. It is essentially spherical and the electrons representing the two negative valences move within this sphere. Their most frequent occurrence is at the corners of a tetrahedron that can be imagined to exist within this sphere. The corners at which these hydrogen ions are found form an angle of 103 to 106° with the oxygen nucleus as the apex. The positive valences of the hydrogen ions are only partially neutralized by the negative valences of the oxygen ion. This leaves a preponderance of positive charge on one side of the water molecule and a negative one on the other. This makes the water molecule a dipole (Fig. 2-1).

Water molecules do not exist individually. The hydrogen in the water serves as a connecting link from one molecule to the other. This *hydrogen bonding* is the sharing of one hydrogen atom by two strongly electronegative atoms, when the hydrogen largely belongs to one of them. The result of hydrogen bonding in the case of water is the formation of a hexagonal lattice structure of many molecules held tightly together. Water can therefore be regarded as a giant polymer of hydrogen-bonded water molecules. This situation prevails in ice, where hydrogen bonding is complete. As water melts and as it becomes warmer, more and more of the hydrogen bonds are broken and the lattice work begins to collapse allowing the water molecules to become

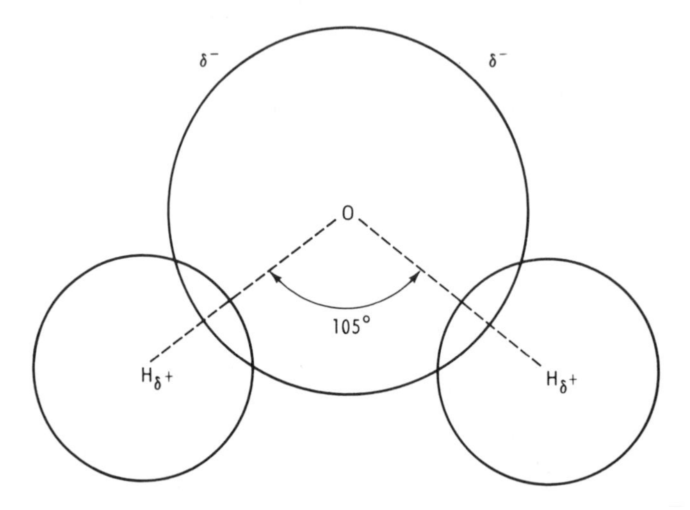

Fig. 2-1 A water molecule: Schematic two-dimensional representation.

more tightly packed, as they are free to follow the van der Waals' forces of attraction. Some hydrogen bonding exists even in the vapor state.

Table 2-1 gives an approximate picture of the effect of temperature on the degree of hydrogen bonding of water.

Hydrogen bonding holds the molecules of water so firmly together that melting point, boiling point, dielectric constant, specific heat, and the viscosity of water are unusually high compared with other compounds of similar and even larger molecular weight. It is because of hydrogen bonding that water is a liquid and not a gas at normal temperatures.

Water at the surface of clay minerals has a structure similar to that of ice. This structure is much more rigid than that of free water and therefore reduces the mobility of this tightly held water.

Water exists in solid, liquid, and gaseous form. Changes from one state to the other are accompanied by energy changes. Water in the solid state has the least kinetic energy; in the liquid state it has more;

Table 2-1 Degree of hydrogen bonding of water

Temperature, °C	*State of water*	*Degree of* H *bonding*
0	Ice	100%
0	Water	85%
40	Water	More than 50%
100	Vapor	More than 10%

water in the gaseous state has the most kinetic energy. The reason for these differences is the difference in the rate of motion of the molecules in these states. Whenever water changes from one state to the other, hydrogen bonds are established or broken. Therefore it requires considerable energy to overcome the transition between these states.

The tendency of the water molecules to polymerize into hexagonal structures is the cause of several interesting phenomena. The most obvious one is the shape of snow crystals. Oxygen, nitrogen, and carbon dioxide gases can penetrate thin polyethylene films but not water vapor. Since the water molecule is the smallest of these four, it should pass the easiest through the tiny pores of the film. It evidently is polymerized even in gaseous condition. Near a solid surface the structure of water appears to be more ordered than in a body of water. Water molecules are attracted to the surface, causing great viscosity in very fine pores. The density of water near solid surfaces is smaller than that of free water. The water in clay-water systems has been found to have an average density between 0.97 and 0.98 g/cc. On the other hand, dissolved ions break hydrogen bonds and disrupt the water structure to cause closer packing and an increased density. While the water structure is a relatively open one, it is firmly held together. Water sticks to itself with great energy. This property is called *cohesion*. The tensile strength of a liquid is a quantitative expression of the energy required to rupture a bar of the liquid. Although an exact measurement of this property of water has not been found possible, it is very large, exceeding 100 atmospheres. Water attaches itself to surfaces of many substances. This is *adhesion*.

ENERGY REQUIRED FOR TRANSFORMATION OF WATER FROM ONE STATE TO ANOTHER

Heat of vaporization Evaporation is an endothermic reaction, a reaction requiring heat. About 580 cal (gram calories) per gram of water are absorbed in evaporation. This is the reason for the considerable cooling effect of evaporation. Heat of evaporation of water is very high compared to that of other liquids. The reason is that in evaporation many hydrogen bonds have to be broken. To evaporate a layer of 1 cm of water covering 1 hectare (about the same as 1 acre-inch), 58 million kcal of heat energy are required; this is about equivalent to the energy of 10 tons of coal or the amount needed to heat a house in the northern United States for one winter. Moist soil stays cool because of evaporation. Condensation, being the opposite of evaporation, is an exothermic reaction. It releases the same amount of energy that is absorbed by evaporation. Therefore soil is heated up by condensation. Water is a cushion against both rising and falling soil temperatures.

Heat of fusion of water Freezing is an exothermic reaction; heat is evolved, about 80 cal (gram calories) per gram. Thawing is an endothermic reaction; heat is absorbed. The amount of heat absorbed in thawing is the same as the amount of heat evolved in freezing. This means that water in the soil protects itself from freezing to great depths in temperate climates because a great deal of heat has to be removed in the freezing of a wet soil.

Snow that is partially melted (wet snow) has a smaller heat of fusion (heat of thawing) than completely crystallized snow. Warm winds or rains will readily melt it. By determining the *thermal value* of the snow, flood hazards can be estimated.

Heat of sublimation Sublimation is the direct change from the solid to the gaseous state. The heat of sublimation is the sum of the heat of melting and the heat of evaporation.

WATER AS A SOLVENT

Water enters in more reactions than any other solvent. Practically everything in the soil is soluble in water, especially if the water is charged with either acids or bases as it frequently is. There are several reasons why water is a good solvent.

1. It is a dipole and consequently can orient itself in such a direction that the negative pole will contact the positive pole of the compound, and the positive pole of the water molecule will contact the negative pole of the compound.
2. The dielectric constant of water is high because it takes much energy to displace the hydrogen bonds. Consequently it will reduce the electric attraction of the dissolved ions. Once they are dissolved, they have difficulty in getting together again.

Table 2-2 Energy transformation of water

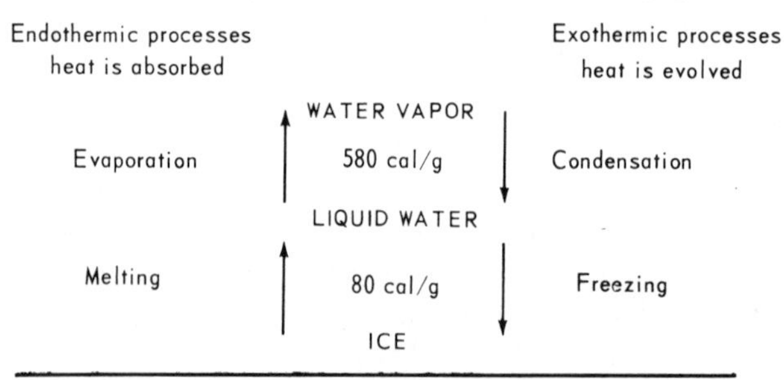

Endothermic processes heat is absorbed		Exothermic processes heat is evolved
	WATER VAPOR	
Evaporation ↑	580 cal/g	↓ Condensation
	LIQUID WATER	
Melting ↑	80 cal/g	↓ Freezing
	ICE	

3. Hydrogen bonding of the water molecules with other substances makes water one of the most reactive compounds. For example: C_2H_5OH is miscible with water.

Solubility measures the strength of stability of a crystal. A salt is soluble in water when the attraction of the ions for water molecules is stronger than the attraction of the ions for each other or, to put it more precisely, when the energy of hydration is greater than the lattice energy.

The solubility of most salts in water increases with temperature. Generally it can be said that the warmer the water, the faster the salt will be dissolved. The solubility of gases, however, decreases with temperature. The following consideration explains this apparent paradox. The higher the temperature of a substance, the greater is its tendency to disperse. Consequently solids will dissolve more readily in water the hotter they are. On the other hand, gases are already in a high state of dispersion. As they get hotter, they leave the water and disperse in the air. Water can dissolve more CO_2 than O_2 and more O_2 than N_2. Cold rains bring much oxygen into the ground. It would be more correct to say that CO_2 reacts with water to form carbonic acid. This is the reason for the apparent great solubility of CO_2 in water.

The ability of water to dissolve most substances is of greatest importance for soil formation and plant growth.

DENSITY

Most substances decrease in density with increasing temperatures. Above 4°C this is also true of water. Below this temperature, however, the density of water increases with increasing temperature.

If the molecules of water would be laid together, like so many marbles, the density of water would be 1.84 g/ml. As it actually is near 1.0 g/ml, the structure must be considerably looser. As we have seen, this is due to hydrogen bonding that causes the water molecules to take on a lattice-type arrangement with open spaces between the molecules. Since hydrogen bonds are broken with increasing temperature, it would appear that the density of water should increase with temperature. On the other hand, kinetic motion increases with temperature and each molecule takes up a greater space. Consequently there are two opposing factors determining the dependence of water density on temperature: the structure, which depends on the degree of hydrogen bonding, and the intermolecular distance, which depends on the kinetic energy of the molecules. From 0 to 4°C the effect of hydrogen bonding predominates; above this temperature the effect of the increased intermolecular distance on density predominates.

The density changes of water in the liquid form are small within

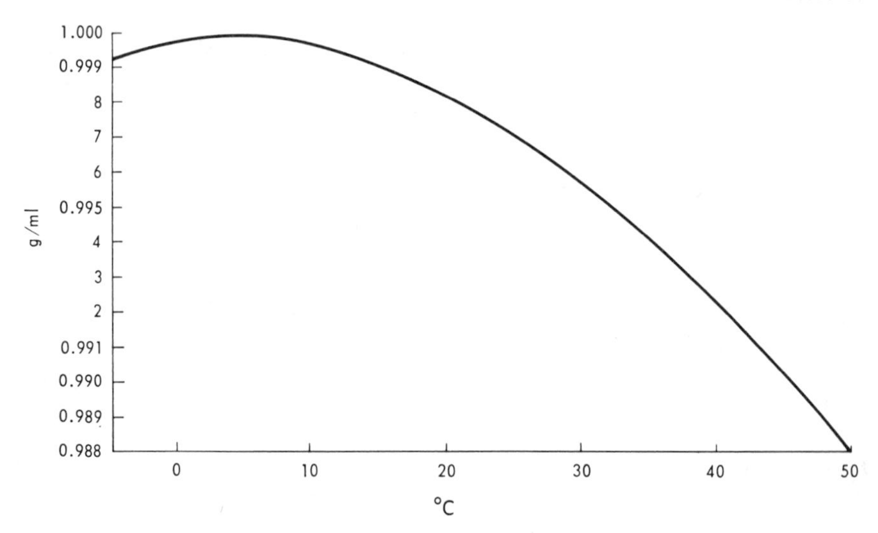

Fig. 2-2 The effect of temperature on the density of water.

the temperature range to which soils are subjected in nature. The difference between the density of liquid water and ice, however, is so large—about 9 percent—that it can greatly affect structure, development, and productivity of the soil (Fig. 2-2). Since hydrogen bonding in ice is complete, its density is considerably lower (0.917 g/cc) than that of water.

As previously stated, water near a solid surface has a structure similar to that of ice. Therefore, the density of water in a clay-water system is smaller than that of pure water. At 25°C it is about 0.97 to 0.98 g/cc. This ice-like structure extends upward to about 60 Å from the clay surface (Low, 1961).

HEAT CAPACITY

By definition, the heat capacity of water is one calorie per gram. This means that 1 cal of heat energy is required to raise the temperature of 1 g of water from 14.5 to 15.5°C. The specific heat of any substance is its heat capacity in relation to that of water at the same temperature. The heat capacities of practically all other substances are lower than that of water.

The specific heat of ice is 0.493 cal/g. The reason that the heat capacity of ice is so much smaller than that of water is that the hydrogen bonds in ice stay intact. In water they are broken as the temperature rises. This requires the extra energy. The specific heat of air is 0.171 cal/g, but specific heat of air on the volume basis is exceedingly small.

The specific heat of most mineral soil particles is about 0.2 cal/g or about 0.5 cal/cc. Therefore the specific heat of 1 cc of soil with 50 percent porosity is 0.25 cal when dry, but 0.75 cal when saturated with water. The specific heat of aqueous solutions is lower than that of water, because in solutions some of the hydrogen bonds have already been broken by the solute. Consequently less energy is required to break hydrogen bonds.

VAPOR PRESSURE

Vapor pressure is the pressure exerted by the vapor of a liquid when in equilibrium with the liquid. It might be called the tendency to evaporate. Vapor pressure increases with temperature (Fig. 2-3). The simplest way to measure vapor pressure of a substance is to introduce a small amount of it into the closed end of a barometer tube and note the

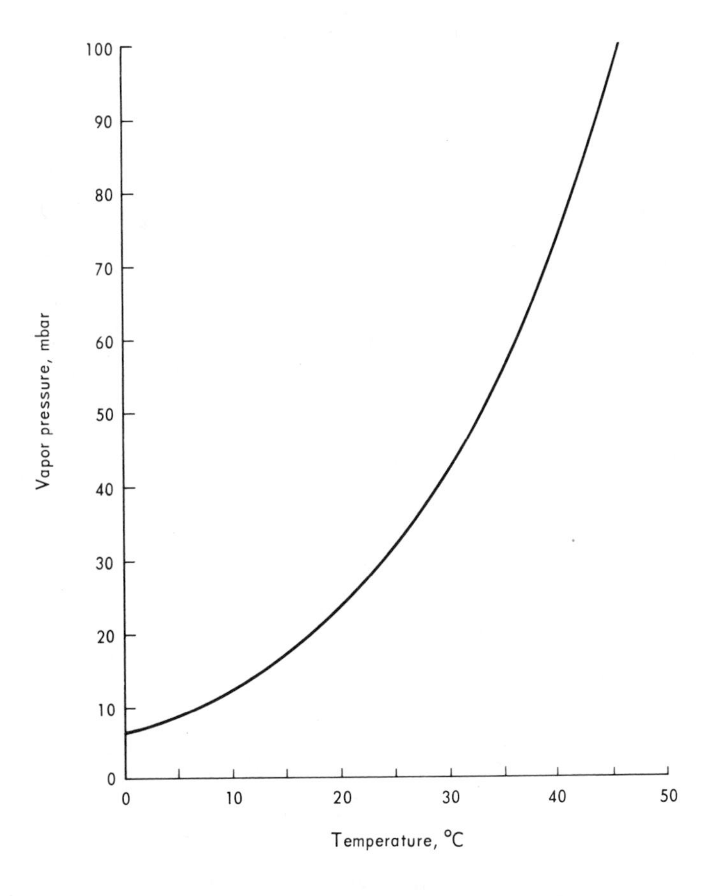

Fig. 2-3 The effect of temperature on the vapor pressure of water.

decrease in the height of the barometer. The vapor pressure of water is very low for a liquid of such small molecular weight.

The vapor pressure of ice and of liquid water at 0°C is the same.

The vapor pressure of boiling water equals 1 atmosphere or the pressure of the system.

Vapor pressure of an aqueous solution is lower than that of pure water because the solute reduces the partial molar free energy of the water. Vapor pressure of soil water is affected both by the solutes and by the soil itself, as well as by the temperature.

Vapor pressure of very small droplets is very high, as it is also a function of the curvature of the surface. The molecules at the surface of such droplets are bonded to fewer neighboring water molecules than if they would be lying at the surface of a flat meniscus. Therefore the

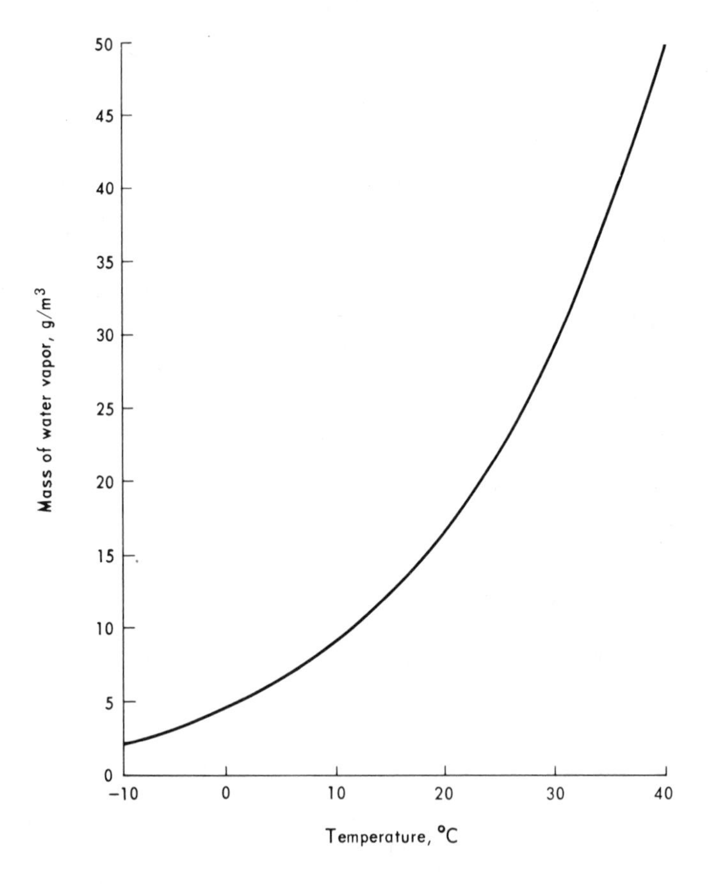

Fig. 2-4 The mass of water vapor in saturated air depends on temperature.

initiation of drop formation from saturated air requires a special stimulus. Consequently air can get supersaturated with water. Dust particles—mostly clay—frequently form nuclei for raindrops.

A given space (volume) at a given temperature and pressure can contain a definite amount of water vapor (Fig. 2-4). This is not affected by the presence of other gases. Frequently not the entire amount of water vapor is present that would saturate the space. The relative saturation of the air with water vapor is called *relative humidity*. This is the amount of water vapor in the atmosphere expressed as a percentage of the total amount the atmosphere could contain at that temperature.

The *dew point* is the temperature at which condensation of the water vapor present in the air takes place.

The principle of the psychrometer which is used to measure relative humidity is based on the fact that the rate of evaporation of water—and therefore the amount of energy required—increases as the atmosphere gets drier. The wet bulb temperature is therefore a function of the energy required to evaporate water in a given medium, while the dry bulb registers the ambient temperature.

BOILING POINT

The normal boiling point of a liquid is the temperature at which its maximum or "saturated" vapor pressure is equal to the normal atmospheric pressure: 760 mm of mercury. The greater the pressure of the system, the higher is the boiling point.

The presence of hydrogen bonds causes the need for much energy to break apart the water polymers. Extra heat has to be supplied for this purpose. Hence the boiling point of water is higher than it would be otherwise for a liquid of so small a molecular weight.

If a solid is dissolved in the water or if another less volatile liquid is mixed with it, the boiling point is raised. The boiling point of water is raised by 0.52°C for each gram molecular weight dissolved in 1 liter of water.

Table 2-3 The boiling point of water changes with temperature

Pressure, mm Hg	*Temperature, °C*
76	46.1
380	81.7
700	97.1
760	100.0
800	101.4

FREEZING POINT

The freezing point of a substance is the temperature at which its solid and liquid forms are in equilibrium. It is the same as the melting point. The normal freezing point of water has been used to establish the zero point on the centigrade temperature scale.

The freezing point of water is lowered by about 1.85°/g mole of solute in 1,000 g of water. This figure refers to the "ultimate" particles of the solute, i.e., the molecules of a nondissociating substance, or the sum of the dissociated ions and the nondissociated molecules of a dissociating compound. The reason that a solute causes a depression of the freezing point is that the molecules or the ions of the solute interfere with the formation of the ice structure. This indicates why the freezing point of the soil solution at the wilting point is below 0°C (−1.22°C). The freezing point decreases as pressure increases (ice under skates melts). The reason that increased pressure prevents solidification (freezing) of water is that the distance between water molecules increases when ice molecules form. And pressure inhibits such expansion.

VISCOSITY

Viscosity of a liquid or a gas is its internal friction that brings about resistance to flow. Viscosity of water, as of most liquids, changes greatly with temperature. Viscosity is measured in poises. A poise is a dyne-second per square centimeter; one-hundredth of a poise is called a centipoise. The viscosity of water at 20°C is almost exactly 1 centipoise. The viscosity of water is very high for a liquid of so low a molecular weight. Hydrogen bonding is the cause of this. The viscosity of ice (glaciers) is 12×10^{13} poises.

Viscosity is an important factor in infiltration and percolation of water. The high rates of infiltration in the tropics, while mostly due to the type of clay and its stable aggregation, are also a result of the low

Table 2-4 The viscosity of water changes with temperature

Temperature, °C	*Viscosity of water*
0	0.01792 poise or 1.000 specific viscosity
20	0.01005 poise or 0.561 specific viscosity
50	0.00549 poise or 0.307 specific viscosity
100	0.00284 poise or 0.158 specific viscosity

viscosity of warm water. Near clay surfaces the viscosity of water is very high, so that its movement is slowed down or even ceases completely. Knowledge of the viscosity of water is needed for the sedimentation techniques used for mechanical analysis of soils.

SURFACE TENSION

Fluid surfaces exhibit certain features resembling the properties of a stretched elastic membrane; hence the name *surface tension.* When dissimilar substances make contact at an interface, the inequalities of molecular attraction (cohesion) tend to change the shape of the interface until, in accordance with the "least energy principle," the potential energy of the whole molecular system attains a minimum value.

Surface tension is a force pulling inward at the surface of a liquid, tending to make the surface area as small as possible. It is due to the unbalancing of the forces of attraction between the molecules, the surface molecules having only the relatively distant gas molecules to pull them outward.

To make it easier to visualize and calculate the nature and dimensions of surface tension, consider the force that holds a soap bubble together. At an imagined central plane of a soap bubble the upper half is held to the lower half by surface tension. Surface tension acts all along the length of the circle that forms the contact between the upper and the lower half. As there is a surface on the outside as well as on the inside of the bubble, the length over which surface tension acts is twice as long as the circumference of the bubble:

$$4\pi rs$$

where s = surface tension.
(The dimensions of surface tension are energy per area, ergs/cm^2, or force per distance, dynes/cm.) The pressure on the inside of the bubble (P_B) is greater than the atmospheric pressure (P_A) on the outside of the bubble. Consequently the difference of the pressures ($P_B - P_A$) acting over the entire area of the cross section of the bubble (πr^2) is the force counteracting the inward pull of surface tension. At equilibrium the two forces must be equal.

$$4\pi rs = (P_B - P_A)\pi r^2$$

This becomes

$$4s = (P_B - P_A)r$$

or

$$P_B - P_A = \frac{4s}{r}$$

This shows that the bigger the radius of the bubble, the smaller is the difference in pressure, or—as atmospheric pressure can be taken as a constant—the smaller is the inside pressure. This can also be stated: The greater the curvature of the bubble, the greater is the inside pressure.

The surface tension of water does not change greatly with temperature—within the temperature limits normally occurring in soil.

CAPILLARITY

One of the important phenomena of surface tension in soil is capillarity, the attraction of water into openings of the approximate diameter of a hair (Latin for "hair" is *capilla*).

Capillarity or "wick action" depends both on the cohesion of the liquid and on the adhesion of the liquid to the solid walls. The height to which the liquid rises is determined by the surface tension and the weight of the liquid column.

Water is not pulled up by capillarity; it is pushed up by a pressure difference. There is always a pressure difference across a curved air-water interface with the pressure under the concave meniscus smaller than the pressure under the plane meniscus of the same liquid at the same height. This causes the water to rise in the capillary (Fig. 2-5). All the sides of the capillary do is to attract the water so that a concave

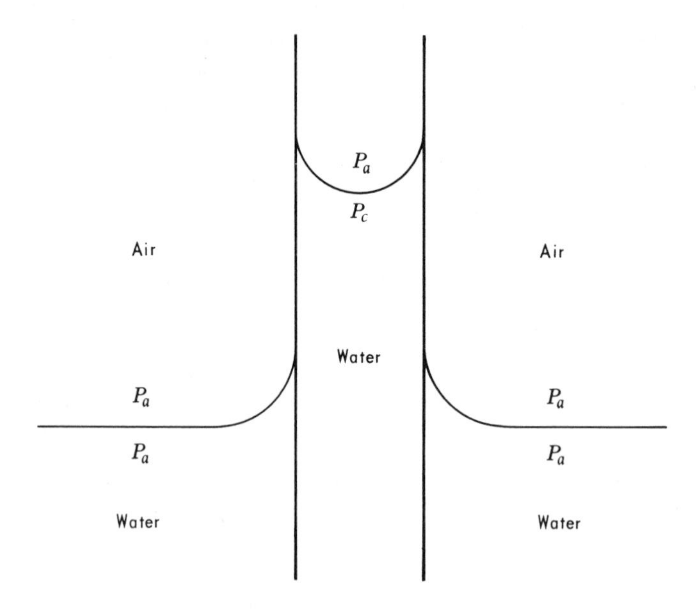

Fig. 2-5 Capillary rise of water: P_a, atmospheric pressure; P_c, pressure of water at the concave meniscus; $P_c < P_a$.

meniscus is formed as a result of surface tension. Such capillary attraction or capillary lift occurs only in a medium in which the walls of the capillary have more attraction for the liquid than the molecules of the liquid have for each other. This is the case for water in most soils and in clean glass. In the case of a convex meniscus the situation is reversed. Water in a wax capillary forms a convex meniscus because the water is not attracted by the wax and consequently the pressure difference at the air-water interface drives the water down.

An equation that permits the calculation of the height of rise of a liquid in a cylindrical capillary can be derived in the following manner.

At equilibrium the force tending to move the liquid downward must be equal but opposite to the force tending to move the liquid upward.

The downward force is the product of

h, height of the liquid above its free surface (dimension: L),
D, density of the liquid (dimension: ML^{-3}),
g, acceleration due to gravity (dimension LT^{-2}), and
πr^2, cross-sectional area of the cylindrical capillary (dimension: L^2).

The upward force is the product of

s, surface tension of the liquid (dimension: ML^{-2}),
$2\pi r$, line of contact between the liquid and the tube (dimension: L), and
$\cos \alpha$, cosine of the angle of contact (this is dimensionless, since it is a ratio). It is a function of the curvature of the meniscus.

Putting these items on both sides of an equation, we get

$$hDg\pi r^2 = s2\pi r \cos \alpha \quad (1)$$

The dimensions on both sides are those of force:

$$L \times ML^{-3} \times LT^{-2} \times L^2 = MT^{-2} \times L$$

$$MLT^{-2} = MLT^{-2}$$

Canceling out π and one r, Eq. (1) becomes

$$hDgr = 2s \cos \alpha$$

Solving for h,

$$h = \frac{2s \cos \alpha}{Dgr}$$

For water the angle of contact is so small that $\cos \alpha$ is practically

1 and the equation becomes

$$h = \frac{2s}{Dgr}$$

or

$$h = \frac{4s}{Dgd}$$

where d is the diameter of the capillary.

Assuming for water at 20°C

$$s = 72.75 \text{ dynes/cm}$$

$$D = 0.998 \text{ g/cc}$$

$$g = 981 \text{ ergs/g-cm}$$

$$h = \frac{0.297}{d} \tag{2}$$

(Both h and d are in centimeters.) For approximation this can be rounded off to

$$h = \frac{0.3}{d}$$

Solving this equation for d, we get

$$d = \frac{0.3}{h} \tag{3}$$

Equation (2) serves to calculate the height of water rise in a soil for which the size of the largest effective pore space is known.

Equation (3) serves to calculate the diameter of the largest effective pore of a soil for which the capillary rise is known.

The relationship between the diameter of soil pores and the height to which water rises above a free-water surface is shown in Fig. 2-6.

Since the density and the surface tension of water change with temperature, the capillary-rise factor changes also. Table 2-5 indicates

Table 2-5 Effect of temperature on capillary rise

Temperature, °C	*dh, the capillary-rise factor*	*Rise of water in a glass capillary of* 0.06 *mm in diameter, cm*
0	0.3083	51.38
10	0.3027	50.45
20	0.2972	49.52
30	0.2915	48.58
40	0.2858	47.64

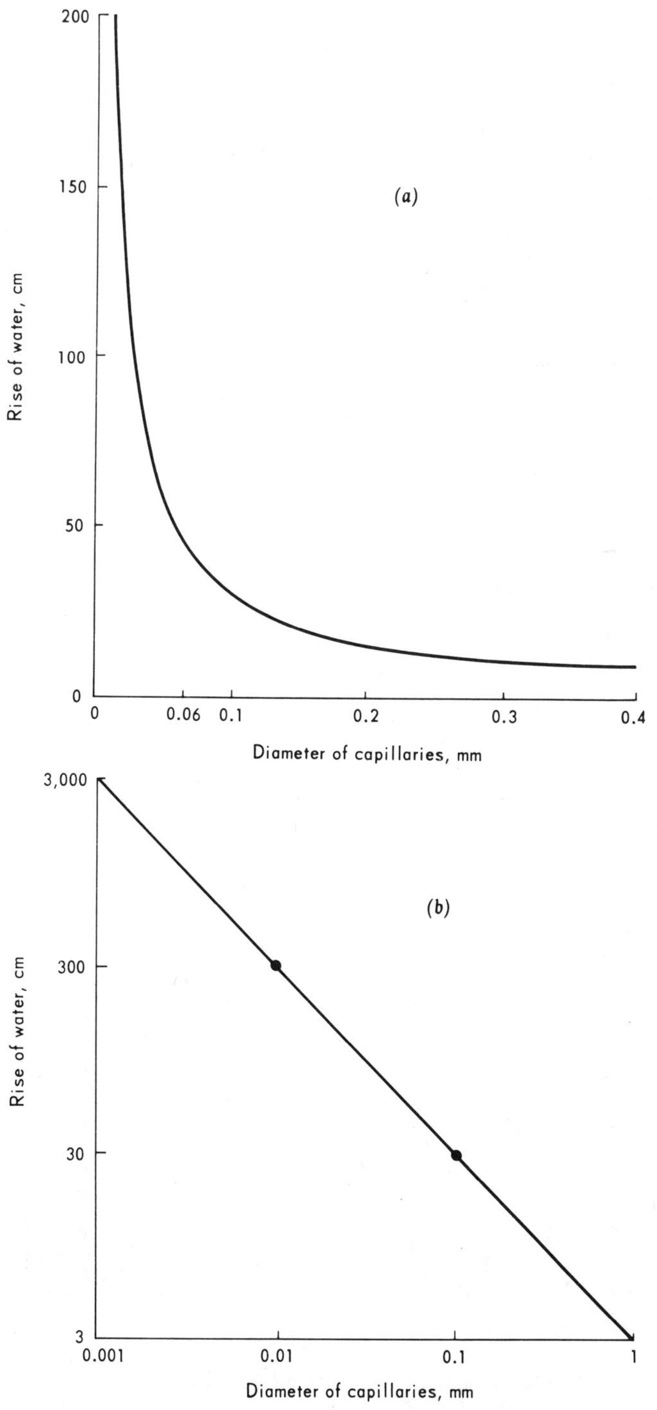

Fig. 2-6 Relation between the diameter of capillaries and the rise of water in soil: (*a*) linear presentation, (*b*) logarithmic presentation.

this relationship. It shows that the height to which water is lifted in a capillary decreases with increasing temperature. To find the capillary-rise factor (dh) for the temperature range occurring in soils, the following equation can be used: $dh = 0.3083 - 0.00056°C$. In these calculations acceleration due to gravity is assumed to be constant: 980.665 cm/sec^2. The relation between the diameter of capillaries and the rise of water in soil is illustrated in Fig. 2-6*a* and *b*.

Quite obviously there are no tubes of uniform diameter in the soil. The height of capillary climb will therefore depend on the largest opening that the water encounters. Once the soil is saturated with water, and then the water is allowed to drain away, some water will be held by the smaller capillaries, even though the larger openings will be freed of water.

The larger the diameter of the capillaries, the faster is the rate of rise. In fine-textured soil the rise becomes exceedingly slow so that plants seldom gain from the presence of groundwater if its level is 80 cm or more below the roots. This is especially true after the roots have exhausted much of the available water supply, because capillary rise in a slightly moist soil is much slower than in a wet one. As a matter of fact the flow of liquid water ceases at around 1 atmosphere tension (pF 3). In the range between pF 3 and 4.2 roots have to grow toward the water or receive it by vapor movement.

In soils the cross section of the pores changes abruptly from spot

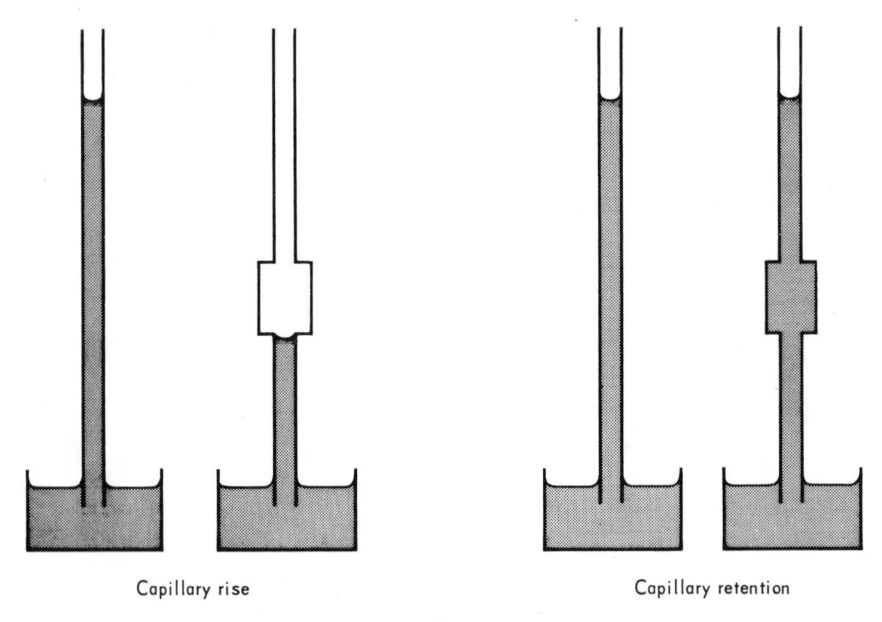

Fig. 2-7 Capillary rise and capillary retention.

Table 2-6 The effect of temperature on some properties of water

Temperature, °C	*Vapor pressure of water, mm Hg*	*Mass of water vapor in saturated air, g/m³*	*Viscosity of water, centi-poises*	*Surface tension of water, dynes/cm*	*Density of water, g/ml*	*Heat capacity of water, cal/g/°C*
−20	0.776	0.892				
−10	1.950	2.154			0.99815	
0	4.579	4.835	1.792	75.6	0.99987	1.00738
4	6.101	6.330	1.567	75.0	1.00000	1.00430
5	6.543	6.761	1.519	74.9	0.99999	1.00368
10	9.209	9.330	1.308	74.2	0.99973	1.00129
15	12.788	12.712	1.140	73.5	0.99913	0.99976
20	17.535	17.118	1.005	72.7	0.99823	0.99883
25	23.756	22.796	0.894	72.0	0.99707	0.99828
30	31.824	30.039	0.801	71.2	0.99567	0.99802
40	55.324	50.5	0.656	69.6	0.99224	0.99804
50	92.51		0.549	67.9	0.98807	0.99854
75	289.1		0.380	63.5	0.97489	1.00143
100	760.0		0.284	58.9	0.95838	1.00697

to spot. The height of capillary rise is therefore limited by the largest diameter. On the other hand, retention of capillary water depends on the diameter of the pore at which there is the contact between air and water, where surface tension is active. Figure 2-7 illustrates this situation schematically.

THE EFFECT OF TEMPERATURE ON THE PROPERTIES OF WATER

The effect of temperature on some of the important properties of water is shown in Tables 2-6 and 2-7.

Although temperature affects all properties of water, it does not do it at the same rate. From 0 to 50°C vapor pressure increases 20-fold, viscosity decreases to one-third, surface tension decreases 10 percent,

Table 2-7 Solubility of gases in water

	Grams of gas dissolved in 100 g of water		
Temperature, °C	*Nitrogen*	*Oxygen*	*Carbon dioxide*
0	0.002942	0.006945	0.3346
50	0.001216	0.002657	0.0761

density decreases 1 percent, specific heat decreases ½ percent, solubility of nitrogen gas in water decreases to one-third, solubility of oxygen gas in water decreases to two-fifths, and solubility of carbon dioxide in water decreases to one forty-fourth.

HYDROLOGY

THE WATER CYCLE

Hydrology is the study of the water cycle in nature. The ever-recurring conversion of ocean water to atmospheric water, to precipitation, to groundwater, to runoff, and to ocean water with its many ramifications is called hydrologic cycle or water cycle. Actually this is not a simple cycle, but a whole system of cycles.

Climatology and meteorology are branches of science that will not be traced here. We are mainly interested in making agriculturally the most of what climate we happen to have. The study of hydrology helps us to understand the amounts of water available to the crops during the various parts of the year and the effects of water on soil and plant.

Table 2-8 shows the hydrologic cycle. Surface storage and inter-

Table 2-8 The hydrologic cycle

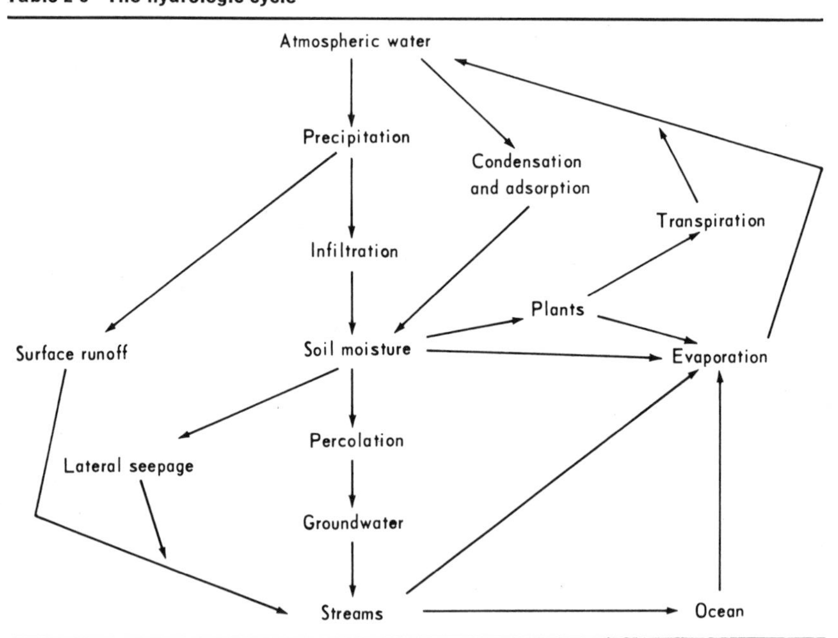

ception are not shown. Also evaporation in falling and from interception and storage is not shown.

THE SOIL-WATER EQUATION

In order to determine the magnitude of the individual steps of the hydrologic cycle as it affects the soil, measurements of various of these items can be made directly by use of rain gages, weighing lysimeters, and direct soil-moisture determinations. For estimation of others the soil-water equation can be used.

Water gain	—	*Water loss*	=	*Water storage* (*This may be positive or negative*)
Precipitation		Runoff		Change in soil-moisture content
rain		Percolation		Interception storage
snow		Evaporation		Surface storage
hail		Transpiration		depression storage
sleet				surface detention
fog (fog drip)				
Condensation				
dew (plants)				
condensation (soil)				
Adsorption				

This equation is applicable to a given soil mass for any period of time studied.

PRECIPITATION

Forms of precipitation Precipitation may occur as rain, fog drip, snow, hail, or sleet. Sleet is a mixture of fine driving rain with small ice particles.

Seasonal distribution There are many types of seasonal distribution of precipitation over the globe, depending on such factors as latitude, altitude, distance from the ocean, and direction of the wind.

In the temperate zone intense rains occur mostly in the summer. The reason for this is that warm air can hold more water and cold fronts in the summer push up the water-laden air into elevations where the temperature is below the dew point. The same temperature difference in winter can only cause the precipitation of much less water, because the mass of water vapor in saturated air is much smaller in low temperatures than in high temperatures (note Table 2-6). Typical distribution of precipitation for several localities is shown in Fig. 2-8.

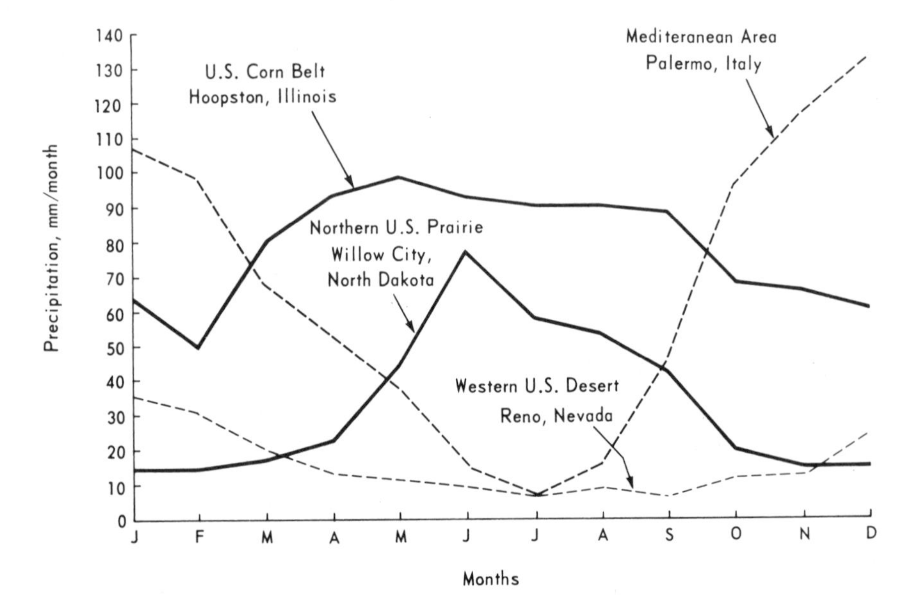

Fig. 2-8 Examples of seasonal distribution of precipitation.

Rain characteristics Rains may be classified according to amount (millimeters), duration (hours), intensity (millimeters per hour), drop size (millimeters in diameter), drop velocity (kilometers per hour), or the energy of impact (kilogram meters per hectare). Precipitation is measured by rain or snow gages. The "Standard" gages merely show the amount of precipitation since the last reading. The "Recording" gages supply a chart that shows the rates as well as the amounts of precipitation. Raindrops can get as large as 6 or 7 mm in diameter. Surface tension of water is not strong enough to hold larger drops against the air resistance encountered. The terminal velocity of the fall of raindrops of more

Table 2-9 Classification of rainfall intensity

Rainfall intensity rate		
mm/hr	*in./hr*	*Descriptive term*
Less than 6	Less than 0.25	Gentle
6–12	0.25–0.50	Medium
12–50	0.50–2.0	Heavy
Over 50	Over 2.0	Severe

than 4 mm in diameter is about 23 km/hr in wind-free air. Very small raindrops are nearly spherical; larger ones are flattened on the underside by air resistance.

Rainfall can be classified according to its intensity as shown in Table 2-9.

If rainfall is to be classified according to its potential erosiveness, both its intensity and its total quantity per storm must be considered. According to the U.S. Weather Bureau, a storm is excessive if it is more intense than $(0.20 + 0.01T)/T \times 60$ in./hr, where T is the duration of the rain in minutes. It appears, however, that a minimum quantity of total rain per storm should be designated, since with the Weather Bureau formula even a storm of only 6 mm (0.25 in.) total would be excessive, if it falls at a rate of 76 mm/hr (3 in./hr) for 5 minutes. Such a small amount of water could hardly cause any runoff and therefore no erosion. It has been suggested to modify the Weather Bureau formula to the extent of stating that a storm of a duration of less than an hour is excessive if its total amount exceeds 0.80 in. (20 mm) (Kohnke and Bertrand, 1959). Wischmeier (1959) has found that the best method of expressing the erosive power of a rainstorm is to multiply its total kinetic energy (in foot-tons per acre) with its maximum 30-minute intensity (in inches per hour). This "rainfall erosion index" takes into account both the detaching and the transporting potential of the rainstorm. It is possible to determine the probable seasonal distribution and the average annual value of the rainfall erosion index for any location for which sufficient precipitation records are available (Wischmeier and Smith, 1965).

INTERCEPTION AND INTERCEPTION STORAGE

Interception is defined as the interference with direct precipitation to the ground by plants or other soil cover. Interception storage is that part of interception that remains in the vegetation until it evaporates.

Precipitation frequently does not reach the soil because it is intercepted by foliage, mulch, or other objects. Some of this water evaporates before it has a chance to drop to the ground. Some is carried down the stem of the plants as *stem flow*. After a small shower in a corn field the soil near the plants is quite wet as a result of the rainwater collected by the corn leaves, while the rest of the land is only slightly damp.

The total amount of water held by interception (interception storage) varies according to the nature of the cover. A good stand of corn can hold about 0.5 mm (0.02 in.). Considerably more water can be held as snow in a dense spruce forest (up to 13 mm). The rate of rainfall and the amounts per storm determine how much of the intercepted water evaporates back into the atmosphere and how much will enter the plant directly and how much will drop or run off to the ground.

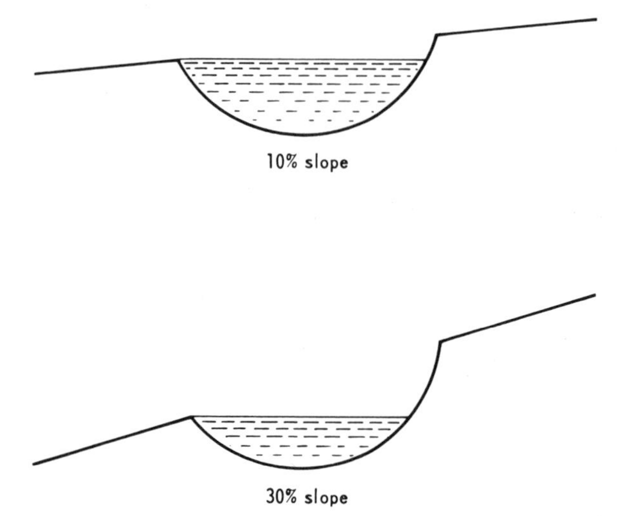

Fig. 2-9 The effect of slope steepness on depression storage: The same depression in the soil retains much more water on a gentle slope than on a steep one.

SURFACE STORAGE

Even after the rainfall exceeds the infiltration capacity of an area, runoff does not occur immediately, because some water is held in small depressions and because a thin layer of water is held on the surface. The holding of water in depressions is called *depression storage* or *surface-water retention.* On a field of gentle slope this may represent a sizable amount, especially if contour cultivation is practiced. On steep slopes there is little depression storage (Fig. 2-9).

The thin layer of water above the soil is held by *surface detention.* It is actually water that is moving downhill, but the rate of movement is slow. There is no definite thickness of the water held by surface detention. The farther it is removed from the soil surface, the less it is slowed down by friction. Water held by surface detention is in *transitory storage.* The time required for the water from the most distant point to reach the watershed outlet is called *time of concentration.*

INFILTRATION, PERCOLATION, AND INTERFLOW

These three expressions refer to the movement of liquid water into and through the soil. Infiltration is the entering of water into the soil. Percolation is the downward movement of water through the soil. It is the same as drainage. Percolation occurs predominantly in downward direction. Interflow is lateral seepage of water in relatively pervious soil above a less pervious layer. Such water reappears on the surface of the soil at a lower elevation. Infiltration rate (millimeters per hour) is the rate at which water enters into the soil. Infiltration capacity (millimeters per hour) is the maximum infiltration rate possible under given conditions.* Infiltration capacity can be calculated from hydrologic data available for natural watersheds.

Infiltration can be considered as the first phase of percolation. Infiltration and percolation are mutually limiting. The infiltration capacity cannot exceed percolation capacity, since no more water can enter the soil than is being removed from the surface. And while the percolation rate cannot exceed the rate at which the water enters the soil, the percolation capacity of any given soil horizon may be considerably greater than the infiltration capacity.

Infiltration capacity usually increases with increasing aeration porosity (large pores) and is therefore correlated with land use. The highest infiltration capacity is normally found in protected woodland; pasture, rotation land, continuous-row crop follow in this order. Burning and pasturing decrease infiltration capacity of woodland. Total infiltration is increased if more time is given to the water to infiltrate by slowing down the rate of runoff or by surface storage.

The infiltration capacity of a soil is greatest at the beginning of a storm. As the rainfall continues, the infiltration capacity is reduced progressively. There are several reasons for this decrease. The large, easily accessible openings of the soil are filled with water. The surface is packed by the impact of the raindrops. Colloids in the soil swell and reduce the pore sizes. Fine material from the surface is washed into the ground, plugging up the pores. The continuous sheet of water above the soil and in the upper layer of the soil makes it difficult for the air in the soil to escape and to make room for further water to enter.

Infiltration capacity is a very dynamic property, depending on changes in moisture content and structure of the soil. The change of infiltration capacity with time during a storm is least in coarse sandy, gravelly, or rocky soil where there is no compound structure to be disturbed. The same is true of muck soils that have an extremely water-stable structure. On the other hand, very dry mucks are hydrophobic (water repellant) and consequently it can happen that infiltration capacities are low at the beginning of a rain. Once the organic compounds are moistened, infiltration capacity increases.

* The Committee on Terminology of the Soil Science Society of America (1956) defines infiltration terms as follows: infiltration velocity—the volume of water moving downward into the soil per unit area per unit time; infiltration rate—the maximum rate at which a soil, in a given condition at a given time, can absorb water.

Table 2-10 The effect of texture and vegetation on infiltration capacity

	Infiltration capacity			
	Vegetated soil		Bare soil	
Texture	*in./hr*	*mm/hr*	*in./hr*	*mm/hr*
Loamy sand	2.0	50	1.0	25
Loam	1.0	25	0.5	13
Silt loam	0.6	15	0.3	8
Clay loam	0.2	5	0.1	3

Latosols (oxisols) with nonswelling clays and a deep profile have a high and rather stable infiltration capacity. The greatest changes in infiltration capacity occur in heavy clay soils of the expanding type. Wide cracks allow almost unlimited infiltration at the beginning of a storm. After these cracks swell shut, infiltration practically ceases.

Infiltration capacities depend on so many circumstances that no valid generalizations can be made. The figures in Table 2-10 are merely to give an idea of the order of magnitude of long-continued infiltration capacities of various soil textures.

Infiltration rates are classified by the U.S. Soil Conservation Service (1951) as shown in Table 2-11.

Percolation occurs when water is under pressure or when the tension is smaller than about ½ atmosphere. The reason is that in soils that are drier than this the water is held with so great a tension that it cannot follow the attraction of gravity.

Table 2-11 Classification of infiltration rates

	Infiltration rate	
Descriptive term	*in./hr*	*mm/hr*
Very rapid	> 10.00	> 254
Rapid	5.00 – 10.00	127 – 254
Moderately rapid	2.50 – 5.00	63 – 127
Moderate	0.80 – 2.50	20 – 63
Moderately slow	0.20 – 0.80	5 – 20
Slow	0.05 – 0.20	1 – 5
Very slow	< 0.05	< 1

Percolation is of great importance in soil development and land management. It removes the soluble salts that would otherwise accumulate in the surface soil—as they do in dry regions that have no, or insufficient, percolation; on the other hand, percolation removes valuable plant nutrients, especially nitrate and calcium.

As evaporation and transpiration use up much of the water that enters the soil, the amount of water that percolates through it decreases with depth. The greatest amount of percolation goes on in the top few centimeters. The high clay layer that is frequently found in the *B* horizon of soils slows down the downward movement of water. Percolation capacity is largely a matter of the structure and texture of the soil. The coarser the texture and the more porous the soil, the greater is its percolation capacity.

The largest amount of percolation occurs in the spring when the frost comes out of the ground. It decreases as evaporation and transpiration increase with increasing temperatures. During the latter part of the most active vegetative period, very little water percolates from the surface to lower depths except in very sandy soils.

Percolation rates have been classified for the use by the U.S. Soil Conservation Service as shown in Table 2-12.

The lysimeter is the classical instrument to measure percolation. The first lysimeter was built around 1680 by De la Hire (1703) in France to study the origin of springs. Many lysimeters are inadequate to determine percolation under natural conditions. Some of the common failings of lysimeters are refilled soil, shallow depth, no runoff provision, no weighing device, and no vegetation. Unfortunately such lysimeters are used to explain physical or chemical phenomena in the soil. Water can leave the lysimeter at the bottom only when it is at zero tension. This results in an accumulation of water just above the outlet. Frequently a stagnant condition results. This is made worse by having no

Table 2-12 Classification of soil permeability rates

	Percolation rate	
Descriptive term	*in./hr*	*mm/hr*
Rapid	> 6.3	> 160
Moderately rapid	2.0 – 6.3	50 – 160
Moderate	0.63 – 2.0	16 – 50
Moderately slow	0.20 – 0.63	5.0 – 16
Slow	0.05 – 0.20	1.25 – 5.0
Very slow	< 0.05	< 1.25

surface runoff provision. By separating the soil in the lysimeter from the strata below, upward movement of water is made impossible. Some of the best lysimeters are those built in 1936–1937 at the Hydrologic Research Station near Coshocton, Ohio. Their features are large area (8.1 m^2), relatively great depth (240 cm), undisturbed soil block (natural structure), runoff provision, weight recorders, and natural land use.

Interflow, as described above, is a form of percolation. The significant feature of interflow is the fact that it does not contribute directly to the groundwater but that it reappears on the surface. Water that has entered the soil may encounter a slowly permeable layer, and a large proportion of the water will therefore accumulate on this layer and continue in an essentially horizontal direction until it either finds a more pervious subsoil or comes to the soil surface at some location farther down the slope. Such lateral seepage occurs frequently and helps in the safe disposal of rainwater except in some cases when it forms pronounced springy spots on hillsides that may form the starting points of gullies and landslides.

The proportion of precipitation water that becomes interflow depends on the relative permeability of the various soil horizons. The greatest amount of interflow occurs in a situation of highly permeable surface soils and slowly permeable subsoils, a condition that can be frequently found in forests.

GROUNDWATER

Groundwater is the tension-free continuous mass of water below the soil surface. It fills all the pores of the material in which it occurs. The surface of the groundwater is called the *groundwater table.* It can be found by drilling a hole in the soil and observing the surface level of the water that will fill up the hole. In some cases two or more groundwater levels exist above each other where extensive layers of impervious materials occur. The upper groundwater masses are called *perched groundwater.* The groundwater level changes with the season. It is high in the spring and sinks during the summer.

The *capillary fringe* of groundwater is held by the soil under tension and is therefore soil moisture and not groundwater. By capillary fringe is meant the water in the layer of soil or subsoil into which groundwater enters due to capillary rise. The effective height of the capillary fringe depends on the pore sizes of the soil. Silt-sized pores give the greatest height. Large pores have a small capillary lift, while the water movement in clay-sized pores is so slow as to be ineffective.

The direct value of groundwater to plants depends on whether the roots can reach the capillary fringe. Whether this be the case or not is determined by the depth of the water table, the rate and extent of capillary rise, the aeration of the soil, and the nature of the crop. Established alfalfa, corn, and other deep-rooting plants can benefit from groundwater in pervious soil, if its level is as deep as three or even more meters.

SURFACE RUNOFF

Runoff from a plot or a small natural watershed occurs when the rate of rainfall exceeds the infiltration capacity. The actual relationship between these three hydrologic factors is complicated by interception storage, depression storage, and surface detention of the water. If considered over longer periods, evaporation and transpiration as well as condensation and adsorption have also to be considered.

The water that runs off from the soil without entering it to any appreciable extent is called *surface runoff* or *overland flow*. Water in continuous streams is also surface runoff. It may originate from overland flow, lateral seepage, and groundwater flow.

Overland flow is usually undesirable because the rainwater bypasses the soil without adding to its productivity and because of the erosion danger. In order to decrease surface runoff, the other "Water losses" (see the soil-water equation) should be increased: infiltration, percolation, transpiration, and evaporation. The amount of surface runoff and of stream flow varies more from year to year than does the amount of water available to plants. In "dry" years a relatively larger percentage of the precipitation enters the soil, while the runoff is restricted.

The following equation shows the factors that determine the amount of surface runoff from a given area for a time period chosen.

Surface runoff = water gains − water losses − water storage

where

Water gains	precipitation
	condensation
	adsorption
Water losses	percolation
	evaporation
	transpiration
Water storage	interception storage
	surface storage
	increase in soil-moisture content

For short-time surface runoff, resulting from an individual rainstorm, the slow moisture changes can be neglected. Hence

Surface runoff = precipitation − storage

COMPOSITION OF RUNOFF AND GROUNDWATER

Surface runoff water is high in soil, clay, organic matter, total nitrogen, and absorbed phosphate; it is low in soluble salts, e.g., calcium, bicarbonate, nitrate, and sulfate. An exception is the surface runoff from saline and alkaline areas.

Percolation water and, consequently, groundwater are high in soluble salts (Ca, Mg, K, HCO_3, SO_4, NO_3, and Cl), depending of course on the nature of the soil through which the water flows. They are low in colloid, organic matter, and phosphate. Generally water that has percolated through soil is potable and sanitary as the soil filters out all solid particles and bacteria and absorbs colors, smells, and many chemicals. Where water has passed through a soil that contains a fair percentage of exchangeable calcium, it has a pH slightly above 7.0, because it is buffered with calcium bicarbonate.

When surface runoff and percolation water are mixed in streams, an approximate idea of the proportion of these two components can be obtained by analyzing the water for one or two key ingredients, e.g., bicarbonate or sulfate. For a more accurate determination the use of tracer elements may be useful.

EVAPORATION

Evaporation is an endothermic process, requiring energy. This energy comes almost exclusively from the sun. The energy needed for evaporation is proportional to the gradient of the vapor pressure of the soil moisture to the vapor pressure of the humidity in the air. The amount of evaporation of soil water into the atmosphere therefore depends on the magnitude of the vapor pressure gradient from the soil surface to the atmosphere just above the ground.

The vapor pressure of the atmosphere near the ground depends on:

Temperature: general air temperature, stability of air, sunshine, wind, color of soil surface, mulch, plant cover.

Relative humidity: temperature, wind near the ground (air change), water vaporized into the air through transpiration and evaporation.

The water vapor pressure at the soil surface depends on:

Moisture content (tension) of the soil.

Temperature of the soil surface: general temperature, sunshine, color of the soil, surface mulch, plant cover.

Moisture supply from below: depending mostly on moisture gradient, moisture content, and type of porosity (conductive capacity).

Whether the rate of evaporation is greater from the soil surface or from open water depends on the circumstances.

If a soil is quite moist and the sun is shining, evaporation from the soil is greater than from an open-water surface, because the soil absorbs more heat than does diaphanous water. Under all other conditions the reverse is true. Soils with a tension of 1 atmosphere or drier at the surface have little sustained evaporation because water conduction is extremely small. Cultivation, by bringing moist soil to the surface, encourages evaporation. Evaporation of water from the soil depends on the presence of water at the surface. Coarse-textured soils have too little capillary lift to supply the surface with water.

TRANSPIRATION

Plants are on liquid diet, taking up nutrients in solution and using water also for photosynthesis, for cooling, and for maintaining turgor in their cells. Plants are very heavy drinkers, their water requirements varying from 200 to 1,000 kg of water per 1 kg of dry matter produced. The water requirements of clovers and alfalfa are high, of small grains intermediate, of corn and soybeans low. The water requirements also vary greatly with the stage of growth of the plants.

The amount of water requirement varies inversely with the site and climatic quality. The water requirement decreases with the increase of the humidity of the atmosphere and with the increase in adaptedness of the site to the growth of the given plant. This is an important fact in considering maximum crop production with the available water.

The plant uses the water most efficiently (its water requirement is the least) when its growing conditions are the best. The water requirement changes constantly with the stage of growth and with the environmental conditions. For this reason, determinations of the rate of transpiration and of the water requirement of plants are not directly applicable in areas of different soil and climatic conditions. Since it is difficult to measure evaporation and transpiration separately, these two forms of water vapor movement are frequently combined in hydrologic calculations and called *evapotranspiration*.

CONDENSATION

Condensation is the change of water from the vapor state to the liquid state. It is brought about by oversaturation of space with water vapor. Condensation occurs when water vapor cools down to the dew point. (The dew point is the temperature to which moist air has to cool to become saturated with the water. Therefore the dew point is the beginning of condensation of the water vapor contained in air.) This may be the result of cooling of air or of air striking a cooler object.

Therefore condensation occurs predominantly at night. The amount of water vapor condensing in soil directly is usually quite small. Condensation on plants, dew, is much more important for agriculture. Dew amounting to 13 mm of water per month has been found on corn and to 33 mm on soybeans in Indiana (Brawand and Kohnke, 1952). Condensation of atmospheric water vapor in the soil occurs when the dew point of the atmosphere near the ground is above the temperature of the soil at its surface. Therefore condensation in the soil is more apt to occur in late winter and in spring than at other seasons.

The amount of condensation in a given area depends on the moisture content of the air (relative humidity), on the temperature, and on the daily temperature fluctuations. The higher these three are, the greater is the amount of condensation.

ADSORPTION

Adsorption is the attraction of hygroscopic water by a dry object. Like condensation, adsorption is an exothermic process. For adsorption by the soil to occur there must be a water vapor pressure gradient from the atmosphere to the soil or from another part of the soil. The atmosphere must be "wetter" than the soil and the soil must be drier than the hygroscopic point. This means that adsorption can only occur in a soil that "feels dry" (except in saline soils). Therefore adsorption can only be an important source of water where surface soils get very dry.

Since adsorption depends only on a vapor pressure gradient, it can occur against a negative temperature gradient (from cold to warm). More frequently it occurs in the direction from warm to cold. Once the adsorbed water has reached the vapor pressure of the environment, adsorption ceases. Because the soil surface dries out usually during the day and the relative humidity of the air increases at night, adsorption predominantly occurs at night.

The amount of water that the soil receives by condensation and adsorption together can be quite sizable. Research in the Corn Belt indicates that approximately 100 mm are received by the soil through condensation and adsorption during one year (Brawand and Kohnke, 1952).

Adsorption of water by the soil is greatest where there are great daily variations in the water vapor pressure of the air. This occurs, for example, along some coasts, where dry land winds during the day and moist sea winds during the night alternate.

CLIMATE AND MICROCLIMATE

In any consideration of hydrology the climate of the area and the microclimate of the specific location have to be taken into account. Climate is the sum total of the weather, while microclimate refers to the climate near the ground. Temperature and moisture conditions near the ground may be quite different from the situation at an elevation of 2 or 5 m. The microclimate is affected by the soil, but in itself affects the various hydrologic and thermal phenomena in the soil.

The availability of precipitation for plant growth and percolation decreases with increasing temperature of the location, since evaporation and transpiration increase with increasing temperature. It is for this reason that Lang (1915) has suggested to classify the humidity of climates by dividing the average annual precipitation in millimeters by the average annual temperature in degrees centigrade. This classification is shown in Table 2-13.

The Lang system gives a satisfactory classification of the humidity of the climate wherever the winter temperatures are mostly above the freezing point. In the temperate and the cold zones it is best to use an adjustment of the Lang factor. This can be done by replacing in the Lang quotient the average annual temperature by the sum of the average monthly temperatures that are above 0°C, divided by 12. This *biotemperature* was first suggested by Holdridge (1962). It is believed that this adjusted Lang factor is a good method of classifying climatic humidity.

Quite obviously the temperature as such has to be taken into consideration too. Using average annual temperature as the ordinate and the bioclimatic factor as the abscissa, a very descriptive system of climate classification is obtained.

UTILIZATION OF HYDROLOGIC INFORMATION

As water is needed for all agricultural, domestic, and industrial activities, knowledge of the actual amounts of the various steps of the water cycle are useful or even essential for a wide variety of purposes. The most important ones are:

Agricultural and forestry: Land use capability, crop growth, fertilization potential, drainage, irrigation, soil conservation, cultivation,

Table 2-13 Lang's classification of climatic humidity

Lang's precipitation factor, mm pptn./°C	*Description of climate*
0 – 40	Arid
40 – 60	Semiarid
60 – 100	Semihumid
100 – 160	Humid
> 160	Perhumid

mulching, water supply for the many domestic needs, for livestock at the barn, and in the pasture.

Ecology: As water is frequently the determining factor of plant growth, hydrologic information is a necessity in the study of the habitats of native and cultivated plants.

Urban: Drinking water, other home use, sewage disposal, industrial needs, air conditioning, fire protection, area planning.

Rivers, lakes, and reservoirs: Navigation, amount of water in rivers, lakes, and reservoirs, potential for hydroelectric plants, flood forecasts, flood control, silting, recreation.

Highways: Requirements for drainage, sizes of bridges, culverts, and ditches, elevation of road surface above possible flood levels, location of roads.

It must be recognized that practically all water that is used by mankind has to pass from the atmosphere into the soil, and therefore a complete understanding of soil-water relationships is of utmost importance for an efficient use of water.

THE FUNCTIONS OF WATER

Considering the many phenomena of nature in which water plays an important role, we recognize that—from the viewpoint of mankind—the functions of water can be beneficial or detrimental.

Water feeds the plants,
Water can drown plants,
Water helps to create soil and to differentiate it into horizons,
Water helps to destroy soil by erosion and obviates existing horizons by removing their components,
Water brings plant nutrient elements into plant-available form,
Water leaches out plant nutrient elements,
Water supports microbial fauna and flora that make nutrient elements available to plants,
Water can drown beneficial microbial fauna and flora,
Water carries dissolved oxygen into the soil,
Water prevents entrance of air into the soil,
Water keeps soil from getting too cold,
Water keeps soil from getting too hot.

The many seemingly contradictory effects of water point to the need of a full understanding of the functions of water, in order to make the best use of this important resource.

THE ENERGY OF SOIL-MOISTURE RETENTION

THE FORCES INVOLVED

Water exerts its effects on soil and plant as a function of its free energy. The partial molar free energy of soil water is determined by the moisture content, the attraction of the soil for water, the temperature, and the content of dissolved substances. The higher the moisture content, the smaller the attraction of the soil for water, the higher the temperature, and the lower the content of dissolved substances (generally salts), the greater is the free energy of the water. The attraction for water by the soil is largely a matter of surface relationships. Fine-textured soil with large specific surface has a great attraction for water.

Free energy is a measure of the escaping tendency or of the tendency of a system to react or change. The use of a free-energy scale makes it possible to relate directly soil moisture to the water in plants as well as to the moisture in the air. If we set the free energy of water at an ordinary surface of pure water at zero, the free energy of water held by soil is less than zero. Water that is just slightly held by soil, e.g., water filling the larger pores, has a free energy nearly as great as that of an open-water surface; but water that is very tightly held by the soil, e.g., the hygroscopic water, has a small free energy or, in other words, a great negative free energy. The greater the energy with which water is held by the soil, the smaller is the free energy of the water. But regardless of the level of energy with which the water is held by the soil, the free energy of soil water is always "below zero," i.e., negative.

As stated before, not only soil particles have an effect on the free energy of water. Any other substance with which water comes in contact and any substance dissolved in the water exerts an influence on its free energy. The more particles (molecules or ions) are dissolved in the water, the lower is its free energy. This means that pure free water evaporates more readily than an aqueous solution. As the free energy decreases with an increasing concentration of the solute, the osmotic pressure increases.

As both the amount of water in the soil and the amount of solute in the water affect the free energy of the water, a moisture scale based on free energy cannot be a direct expression of the thickness of the water films around the soil particles. It is this only where the concentration of ions in the water is negligibly small. The energy of water tension in a soil depends on the specific surface as well as the structure of the soil and on its solute content.

The forces that keep soil and water together are all based in the last analysis on the attraction between the individual molecules, both between water and soil molecules (adhesion) and among water molecules themselves (cohesion). In the wet range surface tension is the most important force, while in the dry range adsorption is the main factor.

DIMENSIONS AND TERMINOLOGY

The potential or tension of soil water at any given point is equal to the work per unit mass of water that has to be done to change its energy status to that of pure free water. This energy change can be expressed in terms of height differences by which the soil water has to be lowered to reach the energy level of the open-water surface (centimeters), or in terms of the pressure difference between soil water and free water (atmospheres) or as potential, i.e., work per mass (ergs per gram), or as partial molar free energy or as total stress.

In this book the word "tension" will be used to express this condition of soil water unless a more specific term is needed.

Table 2-14 serves to show the relationships between the various terms used in describing soil-moisture conditions. Using M as the symbol of mass, L as the symbol of length, and T as the symbol of time, the relationships between the different physical dimensions can be expressed in an abbreviated way.

As pressure increases below the water surface with depth, it decreases with height above the water surface. The pressure conditions at the water surface are arbitrarily set at zero.

Table 2-14 Physical dimensions*

Area	= length squared	$= L^2$
Volume	= length cubed	$= L^3$
Density	= mass per volume	$= ML^{-3}$
Specific volume	= volume per mass	$= M^{-1}L^3$
Specific surface	= area per volume	$= L^{-1}$
Velocity	= length per unit of time	$= LT^{-1}$
Acceleration	= velocity increase per unit of time	$= LT^{-2}$
Force	= mass times acceleration	$= MLT^{-2}$
Tension	= opposite force	$= MLT^{-2}$
Energy	= force times distance	$= ML^2T^{-2}$
Work	= applied energy	$= ML^2T^{-2}$
Surface tension	= energy per area = force per distance	$= MT^{-2}$
Pressure	= force per area	$= ML^{-1}T^{-2}$
Stress	= negative pressure	$= ML^{-1}T^{-2}$
Power	= work per unit of time	$= ML^2T^{-3}$
Potential	= work per mass	$= L^2T^{-2}$
Flow	= mass per unit of time	$= MT^{-1}$
Viscosity	= flow per distance	$= ML^{-1}T^{-1}$

* Dyne is the cgs unit of force; erg is the cgs unit of energy.

Soil-moisture tension is brought about at the smaller dimensions by surface tension (capillarity), at the higher dimensions by adhesion. The energy with which water is held by soil has been called "capillary potential" (Buckingham, 1907). This term does not apply over the entire moisture range. In wet soil, as long as there is a continuous column of water, it might be called "hydrostatic potential"; in the intermediate range the term "capillary potential" applies. In the dry range the term "hygroscopic potential" might be appropriate. It is suggested, however, to use a term that covers the entire range, such as "soil-moisture potential" or "soil-moisture suction" or merely "tension."

Schofield (1935) has suggested to use the logarithm of this tension and has given this logarithm the symbol pF, an exponential expression of a free-energy difference (based on the height of a water column above free-water level in centimeters). The term pF does not imply any specific mechanism of soil-moisture tension, which may be caused by a variety of forces. It includes hydrostatic, capillary, osmotic, and hygroscopic forces.

pF is the logarithm of a free-energy difference measured on a gravity scale. The pF function is defined as the logarithm to the base 10 of the numerical value of the negative pressure of the soil moisture expressed in centimeters of water.

$$\text{pF} = \log h$$

While such terms as "soil-moisture potential," "soil-moisture tension," "soil suction," and "total stress" are used to refer to the free-energy level of soil water, it must be clearly kept in mind that they refer to the work per unit mass of water that has to be done to bring it to the condition of free pure water, and that the soil-moisture tension may be made up of a matric and a solute component. The matric component represents the suction with which the water is held by the soil particles. The solute component is the osmotic pressure due to dissolved substances in the water. There is no sharp division between these two components in the soil.

Also, temperature has an effect on the free energy of water and therefore affects the soil-moisture tension.

In critically analyzing the concept of pF we realize that this is not an absolute value, as it is based on the height of a water column, and the energy represented by the water column is subject to changes. The density of water changes, if only minutely, with temperature. The gravitational attraction differs with location and especially with altitude. It is doubtful that these two sources of error combined can change the pF value of a given soil-moisture tension to any appreciable extent, but they would change it. Therefore it has been suggested to use bars or millibars* as the basis of expressing soil-moisture tension. Probably the most useful method would be to use the logarithm of millibars, because a logarithmic value comes nearer to expressing soil-moisture-tension relationships than any numerical system, and because the millibar is an absolute value. The resulting figure, expressing the free energy of soil water, is quite similar to the pF values. For practical reasons the use of pF is maintained in this book.

CHANGES OF SOIL PROPERTIES WITH SOIL-MOISTURE TENSION

Feel Regardless of the type, all soils, excluding those of high electrolyte content, feel dry when their moisture potential is above pF 4.5. All soils feel moist in the pF range 2.5 to 4.5 and all soils feel wet below pF 2.5.

Appearance Soil between the oven-dry condition (pF 7) and the hygroscopic point (pF 4.5) appears light colored. It may be powdery or hard depending on texture and structure. Soils that are wetter than the hygroscopic point are dark, since they absorb the light much more than do dry soils. Soils at the field capacity (pF 2.5) or wetter glisten with moisture. Free water can be observed on saturated soils (zero tension).

Vapor pressure Unless there is an appreciable amount of electrolyte present, the vapor pressure of wet and moist soil is practically, though not completely, the same as that of free water. That means that the relative humidity under these circumstances is near 100 percent. As the soil dries out visibly the vapor pressure, and with it the relative humidity, decreases very rapidly until it becomes 0 percent when the soil is oven dry.

The relationship between relative humidity at 20°C and pF is expressed by the equation

$$\text{pF} = 6.5 + \log(2 - \log \text{RH})$$

where RH = relative humidity in percent (Schofield, 1935).

The freezing point The freezing point of soil water in the wet range is very nearly 0°C, but in the moist range the freezing point decreases gradually and in the dry range the freezing point reaches very low tem-

* A bar is an expression of pressure.

$$
\begin{aligned}
1 \text{ bar} &= 1 \times 10^6 \text{ dynes/cm}^2 \\
1 \text{ millibar} &= 1 \times 10^3 \text{ dynes/cm}^2 \\
1 \text{ bar} &= 0.98692 \text{ atmosphere} \\
1.0133 \text{ bars} &= 1 \text{ atmosphere} \\
1 \text{ bar} &= 1033.26 \text{ g/cm}^2 \\
1 \text{ bar} &= 1022.7 \text{ cm of water at } 25^\circ\text{C}
\end{aligned}
$$

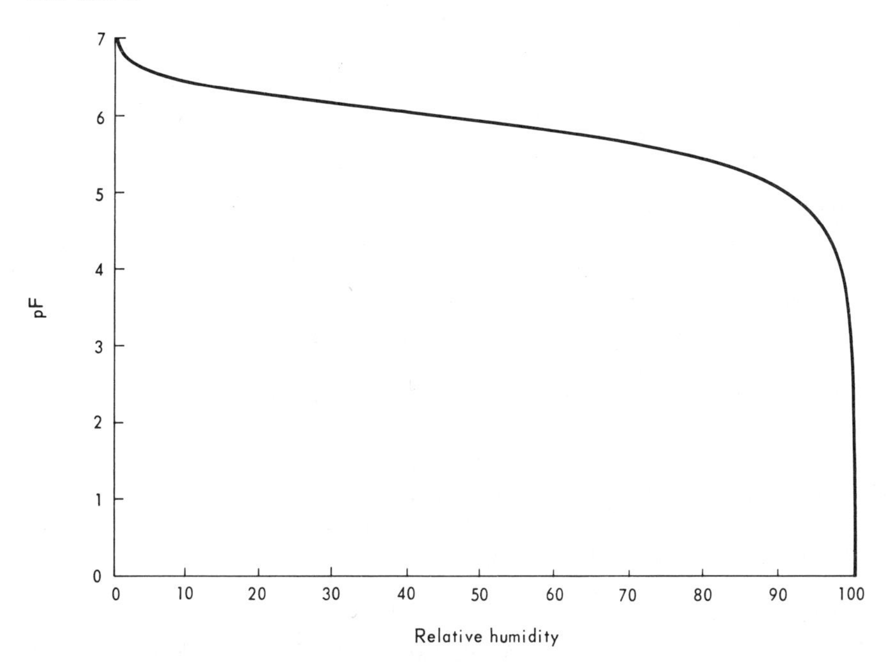

Fig. 2-10 Relations between relative humidity and pF.

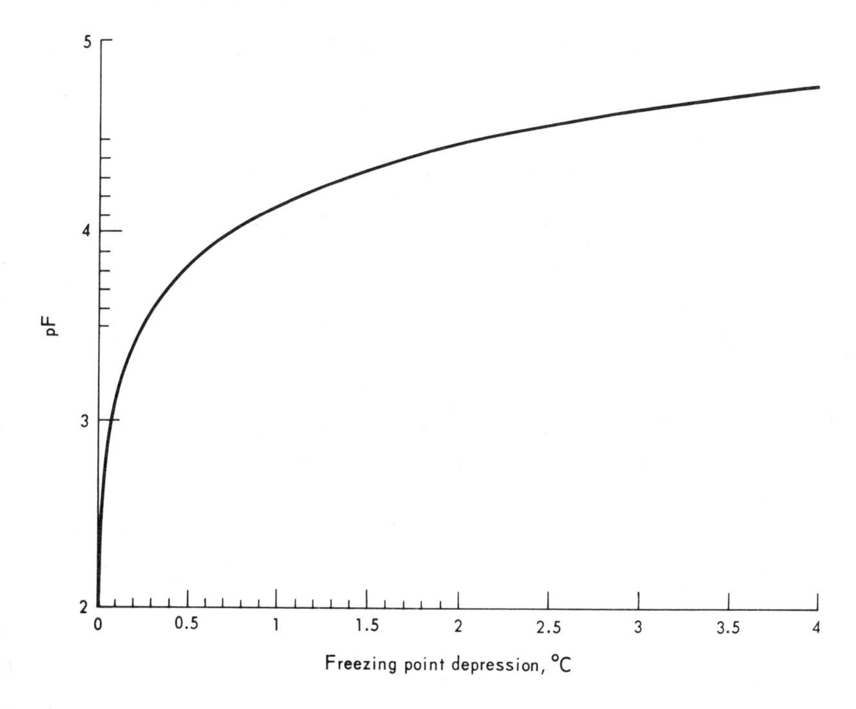

Fig. 2-11 Relation between the freezing point of water and pF.

peratures. Water under pressure freezes at lower temperature than free water. The relationships between freezing-point depression and pF is expressed by the equation

$$pF = 4.1 + \log FPD$$

where FPD = freezing-point depression in degrees centigrade (Schofield 1935).

Aeration Aeration is only approximately related to moisture tension because soil texture and structure will affect aeration also.

Heat of wetting When water is placed in contact with dry soil, the molecules of water are avidly adsorbed by the soil surface and those nearest the solid particles arrange themselves in an ice-like structure. Just as freezing is an exothermic process, wetting of soil by water also produces heat. This is called the *heat of wetting*. Heat of wetting exists essentially only in the dry range (pF 4.5 to 7.0). It decreases as moisture content increases, until it is practically zero at the hygroscopic coefficient.

Heat of wetting is expressed in calories of energy produced per gram of oven-dry soil. The amount of heat of wetting of a soil depends largely upon the amount and the nature of its surfaces. Soils high in organic matter and expanding clay have high heats of wetting. Table 2-15 gives approximate ranges of heats of wetting for soils of different textures.

HYSTERESIS PHENOMENA OF SOIL WATER

When equilibrium is approached in a soil-water system, the moisture content that will be reached depends on the direction of this approach. If a wet soil is exposed to a higher tension, its moisture content at equilibrium will be greater than that of a dry soil exposed to the same tension. In other words the soil moisture content–soil moisture tension curve of a soil will be different depending on whether the soil is being dried or

Table 2-15 Heat of wetting of soils of different textures

Texture	*cal/g*
Sands	<1
Sandy loams	1–2.5
Silt loams	2.5–5
Silty clay loams	5–10
Silty clays	10–15

wetted. For the same tension, the soil will be wetter when it is being dried than when it is being wetted. This can readily be explained in the wet and moist range by the fact that narrow tubes hold water even though they are not connected with a source of water, whereas they cannot absorb water if a large pore hinders water from reaching the narrow tube (Fig. 2-7). In the dry range the attaining of a water equilibrium is an exceedingly slow process. It is questionable whether a true water equilibrium is ever reached in such a soil. It must be assumed that there is a constant rearrangement of water molecules in this range, caused by adsorption and evaporation. This is even true if the temperature remains absolutely constant. Under normal conditions—even in a so-called constant-temperature room—temperatures fluctuate enough to change the free energy of the water. This results in a readjustment of the water molecules on the surface of the soil particles. It must be recalled that in the dry range the layer of water around the soil molecules is less than five molecules thick.

THE VALUE OF USING ENERGY UNITS IN EXPRESSING SOIL MOISTURE

The major advantage of using energy units is that we express soil moisture with reference to the properties of the soil, not merely according to a lifeless correlation like percent. The logarithm of the free energy expresses the moisture content of a soil much more like we would do it according to the way we feel and see the moisture than percent does. Tension varies with moisture content more nearly logarithmically than numerically. Wet soils have low pF values regardless of whether they have 4 or 40 percent water by weight.

This system of expressing moisture content can well be applied to other objects besides soil, e.g., grain, hay, live plants, cloth, sponges, wood.

Reasons for expressing soil-moisture energy conditions in logarithmic form are to simplify terms and to plot a wide variety of moisture conditions on one graph giving the various ranges the proper emphasis. It is not always practical to use logarithmic soil-moisture-tension units. Where data are only of the wet range, logarithmic expressions serve no purpose. If suction is plotted against depth of soil, suction is best plotted in negative depth of water (and both sets of depth data in the same units, e.g., centimeters). A disadvantage of the logarithmic moisture scale is that zero tension cannot be expressed in pF units. There is no logarithm of zero.

The tension of soil water is not only determined by the energy with which the soil holds it, but also by the amount of matter in true solution. Sometimes such solutes increase the osmotic pressure much more than the soil particles themselves do.

This is frequently true in soils that contain appreciable amounts of salts, e.g.:

Saline and alkaline soils of dry region,
Greenhouse soils,
Reclaimed sea-bottom soils, and
Highly fertilized field soils, especially near fertilizer.

It is also true in soils with free acids, e.g.:

Acid bogs and
Acid spoil banks of coal mines, due to oxidation of sulfides to sulfuric acid.

The effect of dissolved matter is to increase the osmotic pressure and to lower the vapor pressure. In this way the pF of a soil solution is much higher than it would be if the soil solution consisted of pure water. In fact a salt solution in a barrel of coarse gravel might have a pF of 4.5 (30.5 atmospheres of osmotic pressure). We can adjust culture solutions to any desired pF by increasing the osmotic pressure. While osmotic pressure affects the vapor pressure and hence the moisture tension, it does not affect the hydrostatic pressure of the water.

From the agricultural point of view we are interested in what soil water does to plants, and it is logical to include both the matric and osmotic forces in expressing the energy status of soil water. pF is therefore defined as the logarithmic expression of the total stress or of the free energy of soil water, not merely as a logarithmic expression of the retention of water by the soil.

The suction method through porous material of determining pore space is designed to measure pore space, not the total tension. It can be used to determine the total tension when the osmotic component at that tension in the soil is very low. Where higher concentrations of solutes are suspected, the osmotic value of the soil solution has to be determined. The sum of the matric and osmotic tension represents the total soil-moisture tension.

METHODS OF MEASURING SOIL-MOISTURE TENSION

Wet range *Suction tensiometer:* A porous cup filled with water is placed in contact with the soil. After equilibrium between the water in the cup and the surrounding soil moisture is established, the tension of the water in the cup is determined with a manometer. Theoretically this method can be used up to 1 atmosphere, because beyond 1 atmosphere the water column breaks. Actually most suction measurements do not go higher than $\frac{3}{4}$ atmosphere, as the pores in the ceramic plates would have to be very fine (0.003 mm in diameter) to prevent air from leaking through, and water movement is extremely slow at the tension of 1 atmosphere.

Tension table: It is possible to determine the amount of moisture a soil holds by matric tension by placing the water-saturated soil on a porous membrane and subjecting the two sides of the membrane to the desired difference in tension. In the wet range this is done by use of the so-called tension table. The upper side of the membrane (usually blotter paper supported by a screen on a plane surface) is exposed to atmospheric pressure. A column of water of the desired height provides the tension on the lower side. The tension table is normally used for tensions up to 1 m of water.

Moist range *Pressure-plate apparatus and pressure-membrane apparatus:* For higher tensions a similar technique is used. In this case, however, the lower side of the membrane is exposed to atmospheric pressure while higher air pressure is created on the upper side. For tensions up to about 1 bar the pressure-plate apparatus is used. The "plate" is made of ceramic material. The pores are sufficiently large to allow a rapid rate of transmission of water. The size of the pores is the reason that air would leak through if the pressure were much above 1 bar.

For pressure up to 20 bars the pressure-membrane apparatus is used. The membrane consists of a sheet of flexible plastic with very fine pores.

Freezing-point-depression method: From saturation to a total tension (matric and osmotic combined) of about 2 or 3 atmospheres, the freezing point of water changes only a little (Fig. 2-11). From 3 to about 25 atmospheres there is a pronounced change of the freezing point. Beyond this level there is so little water in the soil that it becomes practically impossible to determine its freezing point. Therefore the best range to determine total tension by the freezing-point-depression method is from pF 3.5 to 4.4. In cases of strongly saline soils the freezing-point-depression method can be used to determine the total stress to considerably higher levels because of the greater water content present.

Centrifuge: The centrifuge can be used to determine the amount of soil moisture retained against definite centrifugal forces. Briggs and McLane (1907, 1910) have developed a technique in which a wet sample of soil is subjected to a centrifugal force 1,000 times the force of gravity for 40 min. The resultant condition is called the *moisture equivalent.*

In this centrifuge the soil near the outside of the cup is 2 cm farther away from the center of rotation than the soil on the inside. It follows that the soil of one and the same sample is exposed to different centrifugal forces. Moreover, the soil is not used in its natural structure. Therefore the results are only of comparative value. The moisture content at "moisture equivalent" is similar to the moisture content at "field capacity," but there are no definite relationships between the two.

Electrical-conductance methods: Electrical conductance through soil varies with the moisture content, but it is affected by several other factors of which the salt content of the soil, its temperature, and the contact of the electrodes with the soil are the most important ones. Therefore electrical conductance of the soil cannot be used for soil-moisture-tension determinations. Several methods have been developed in which the electrical resistance of porous blocks that are in moisture equilibrium with the surrounding soil is measured. Such blocks were first made by Bouyoucos and Mick (1940). They consisted of two parallel wires embedded in a small block of gypsum. The moisture within the block is always saturated with calcium sulfate. The conductivity varies with the amount of solution in the block. This is assumed to be in equilibrium with the moisture in the soil. Such blocks can be calibrated against the content or the tension of soil moisture and are fairly satisfactory because the saturated calcium sulfate solution overcomes to a certain extent the effect of soluble salts in the soil solution upon electric conductivity. After some time they may lose their calibration, probably because of dissolution and reprecipitation of the calcium sulfate. After about a year in a moist soil the blocks are worn down by solution to such an extent that they cannot be used. To overcome this, a thin plastic coating has been provided for the blocks. The replacement of the two wires by plate-shaped electrodes has also helped to improve the reliability of the readings.

The pores in the gypsum blocks are small. The largest ones are approximately of silt size. These are filled with water when the soil is at field capacity. In this condition the electrical resistance of the block has reached its minimum. Higher moisture contents in the soil do not reduce the resistance further. Therefore gypsum blocks register soil moisture only in the moist range.

In order to estimate soil moisture also in the wet range, material with larger pores, such as nylon or glass fiber, has been used for electrical resistance blocks. Because of the lack of a buffering substance the resistance of these blocks is greatly affected by soluble salts in the soil. Nylon and glass-fiber blocks are therefore of very limited value.

Dry range *Relative humidity:* The soil is allowed to come to equilibrium with an atmosphere of known relative humidity in a constant-temperature room. In this range (dry range) the temperature is of importance, since the amount of water vapor in the atmosphere changes greatly with temperature.

Aqueous sulfuric acid solutions have been used for this purpose. A 3.3% solution of H_2SO_4 in water has an aqueous vapor pressure corresponding to 98 percent relative humidity or pF 4.5. Such a solution has been used to determine the hygroscopic coefficient in soils. The disadvantage of using a dilute solution for this purpose is that its concentration changes during the determination because water is given off to the soil samples or received from them. Therefore the concentration of the H_2SO_4 has to be checked and adjusted. More recently saturated salt solutions are used for setting up definite vapor pressure levels as a means of establishing the relationship between soil-moisture tension and soil-moisture content in the dry range. They have the advantage that the vapor pressure remains the same as long as the solutions are in equilibrium with the solid phase provided that the temperature remains constant. Change of moisture content of soil does not alter the vapor pressure of such a solution as long as part of the solid phase of the salt is left (Table 2-16).

Table 2-17 shows the tension ranges in which the different methods of soil-moisture-tension measurements are effective.

SOIL–MOISTURE CONSTANTS AND CLASSIFICATION OF SOIL MOISTURE

REASONS FOR ESTABLISHING SOIL–MOISTURE CONSTANTS

A knowledge of the amount of water held by the soil at the various tensions is required if we are to calculate the amount of water that is available to plants, the water that can be taken up by the soil before percolation starts, the amount of water that should be used for irrigation, and many other items of hydrologic importance.

Table 2-16 Examples of saturated salt solutions used to obtain definite water vapor tensions (temperature: 25°C)

Salt	*Relative humidity*	*pF*
$CaSO_4$	97.8	4.49
$NH_4H_2PO_4$	93.0	5.00
NH_4Cl	79.3	5.51
$Mg(NO_3)_2$	52.0	5.96
$KC_2H_3O_2$	19.9	6.36

Table 2-17 Ranges of methods of determining soil-moisture tension

Method	*Approximate range of application, pF*	*Type of tension measured*
Tension table	0–2	Matric
Field tensiometer	0–2.8	Matric
Pressure plate	0–2.8	Matric
Centrifuge	2–2.8	Matric
Gypsum electric resistance blocks	2.5–4.1	Matric
Pressure membrane	3–4.2	Matric
Freezing-point depression	3.5–4.4	Matric and osmotic
Vapor pressure	4.4–7	Matric and osmotic

The relationship between the amount of soil water and the energy with which it is held is a continuous function without any sharp breaks. For reasons of practical use and for the determination and tabulation of soil-moisture data, it is necessary to select definite tension levels as reference points.

Since we prefer to express soil-moisture constants in terms of the energy with which the water is held at that particular moisture content, only soil-moisture constants that are reasonably "equipotential" in nature are used.

SATURATION

A soil whose pores are completely filled with water is said to be *saturated.* Generally this implies that the water in the soil is at zero tension, and this condition of saturation is implied in this book. It is possible, however, that the water in a fine-textured soil is under a small tension but air cannot enter it because the menisci at the surface of the soil are strongly curved. On the other hand, soil that is actually flooded normally contains blocked air pores and therefore is not saturated in the strict sense. For practical purposes it is probably best to define a soil as saturated when its water is at zero tension and the majority of its pores are filled with water. This condition occurs much more frequently in nature than complete saturation.

AERATION–POROSITY LIMIT

Aeration porosity of a soil is defined as that part of the pore space—expressed in volume percent—that is freed of water by a tension of water column of a height of 50 cm. Consequently the *aeration-porosity limit* is the soil-moisture tension brought about by a water column of 50 cm in height. This corresponds to a pF of 1.7 or a tension of $\frac{1}{20}$ atmosphere. Aeration porosity is the same as noncapillary porosity. Pores up to a diameter of 0.06 mm are emptied of water. This is about the limit where percolating water begins to slow down considerably.

Tiles placed at a depth of 100 cm can remove the water from the aeration pores within approximately the upper 50 cm of soil, or the zone of the majority of the roots of crops grown on tiled land.

FIELD CAPACITY AND THE 1/3-BAR PERCENTAGE

After a thoroughly wetted soil has drained several days, it reaches a relatively stable moisture condition. This is called *field capacity*. If we postulate that the soil has uniform texture and structure and a deep groundwater level, field capacity could be assumed to occur at a fairly close range of soil-moisture tension. Since the tension exerted on the soil moisture under such conditions is that of a continuous water column of over 10 m it might be assumed that the tension is approximately that of 1 atmosphere. This, however, is not the case even under ideal conditions of a deep uniform soil. One reason is that, before a tension of 1 atmosphere is reached, the water column breaks and becomes discontinuous. The tension at which the water column becomes discontinuous is not the same in all soils. The coarser the soil texture, the lower is the tension at which the column breaks. In silt loams this occurs between $\frac{1}{3}$ and $\frac{1}{2}$ atmosphere. Another reason is that the movement of soil water becomes extremely slow when the tension exceeds $\frac{1}{2}$ atmosphere.

Another factor prevents field capacity from being an equipotential characteristic of the soil. Frequently a less pervious horizon or a groundwater table less than 5 m deep reduces the tension on the water in the surface soil. This results in a soil wetter than corresponds to $\frac{1}{2}$-atmosphere tension.

For these reasons the concept of field capacity is generally abandoned by soil scientists. Instead the *$\frac{1}{3}$-bar percentage* is frequently used to designate the wet limit of the range of plant-available water under general field conditions. The $\frac{1}{3}$-bar percentage is admittedly only an approximate expression of this wet limit, but it is an equipotential characteristic and permits a direct comparison from soil to soil. One-third bar of tension corresponds to a pF of 2.53.

BEST TILLAGE RANGE

Soil should only be tilled when its structure is improved by such an operation. When the soil is too wet it is puddled by tillage. When it is too dry it has such a high degree of cohesion that it breaks into large clods and powder, so that its structure suffers. The soil has to have enough consistence to maintain its small aggregates and to hold up the tillage implements. On the other hand, it should be friable enough to break at the natural cleavage planes between the aggregates.

This very important soil-moisture condition is estimated to fall between pF 2.8 and 4.4. Sands have probably a wider range, because structure in sands is due largely to texture, while heavy soils in poor tilth may have a narrower range. Organic soils can be tilled when they are considerably wetter than pF 2.8 because they have a stable structure. It is probable that different tillage implements require somewhat different soil-moisture ranges for optimum operation. The pF of 2.8 as the wet limit of the best tillage range indicates that in addition to drainage some water has to be lost by evaporation or transpiration before it is fit to be tilled.

WILTING POINT

The wilting point may be defined as the soil-moisture condition at which the ease of release of water to the plant roots is just barely too small to counterbalance the transpiration losses. For purposes of specific definition it is postulated that the atmosphere is saturated with water vapor (Briggs and Schantz, 1912). The wilting point, also called "wilting coefficient" and "permanent wilting percentage," is expressed as moisture percentage by weight. It is one of the most useful soil-moisture constants, and one that can be determined with rather great accuracy.

Richards and Weaver (1943) found the wilting point for the majority of soils investigated to occur at tensions somewhat below 15 atmospheres, while the wilting point of the soils studied by Robertson and Kohnke (1946) averaged 13.6 atmospheres. The 15-atmosphere percentage has been adopted by many soil scientists for, or in place of, the wilting point. More recently, the 15-bar percentage has been used. This corresponds to 14.8 atmospheres or pF 4.18. Although there exists a considerable difference between a tension of 13.6 and 15 atmospheres, the difference in the amount of water in the soil at these two levels is only small.

The wilting point is affected by both the plant species and the stage of growth, but falls for mesophytic plants between 10 and 25 atmospheres. At the wilting point there is a layer of about five or six molecules of water around the soil particles. This corresponds to approximately 20 Å. Since this is an exceedingly thin layer, the moisture content at the wilting point can be determined just as well on a sieved soil sample as on a soil in its natural structure. Conditions of macrostructure are of little influence upon soil-moisture-tension relations at this energy level. On the other hand, the solute content of the soil solution is of great influence upon the permanent wilting percentage. In applying the concept of the wilting point to practical field work it must be kept in mind that normally the roots of a single plant are located in several soil levels that may differ in moisture stress. A plant in a soil that is dry at the surface can thrive, if moisture of lower stress is available farther down.

HYGROSCOPIC COEFFICIENT

Hygroscopicity is the ability of a body to adsorb moisture from the atmosphere. Originally the hygroscopic coefficient of soils was meant to be the condition at which this adsorption is completely satisfied in an atmosphere saturated with water vapor.

There is no practical way to determine the hygroscopic coefficient at 100 percent relative humidity. Soil water at 100 percent relative humidity has zero tension; in other words, the soil is saturated with water. Therefore the hygroscopic point of soils has been set as the moisture tension in which the soil water is in equilibrium with an atmosphere of 98 percent water vapor saturation (98 percent relative humidity). This corresponds to pF 4.5 or 30.5 atmospheres. Ninety-eight percent relative humidity is equivalent to the vapor pressure above a 3.3% aqueous solution of H_2SO_4 in a closed system or above a saturated solution of $CaSO_4$.

The hygroscopic coefficient is the boundary between moist-appearing and dry-appearing soil, and the boundary between soil that exhibits the phenomenon of heat of wetting and soil that does not. None of these boundaries are sharp, and the selection of pF 4.5 is somewhat arbitrary, although essentially sound. The amount of water that a soil holds at the hygroscopic coefficient depends largely upon the amount and type of its specific surface and upon the nature of the exchangeable ions and the presence of free electrolytes. Soils high in expanding clay and in organic matter have a high hygroscopic coefficient.

OVEN DRY

Soil is considered to be *oven dry* when it has reached equilibrium with the vapor pressure of an oven at 105°C. The oven-dry condition corresponds to a relative humidity of approximately 0 percent or a pF near 7.

Depending upon the vapor pressure in the oven, the pF will be between 6.8 and 7.0. This in turn depends upon the temperature and moisture conditions in the room surrounding the oven. The lower the outside temperature and the lower the relative humidity surrounding the oven, the lower the vapor pressure in the oven. These differences are only small. The oven-dry condition of the soil is used as the basis for calculation of the content of the soil of moisture and all other components.

Soil moisture can have higher pF values than 7.0, but most soils give off only little water when they are heated beyond 105°C, until the temperature is high enough to drive off water of constitution. When this has been done, the soil has changed its identity irreversibly.

OTHER SOIL-MOISTURE TERMS

Other terms used to designate certain ranges of soil water are "water-holding capacity," "available water," and "air-dry soil." "Water-holding capacity," or "moisture-holding capacity," as it is also called, is a popular term and may carry a meaning where the tension limits are designated. This, however, is seldom done. Sometimes field capacity is used as the wet limit of the water-holding capacity and sometimes zero tension. Quite obviously this has to be stated, if the term should have any value. It is also not always clear whether the wilting point or the oven-dry condition is meant to be the dry limit. For these reasons it is suggested that the term "water-holding capacity" not be used, except as a general descriptive term.

The situation is quite different with "available water" or "plant-available water." This is usually meant to signify the water held by a soil between ⅓-bar percentage and the wilting point. Of course, it must be made clear whether it means all of the water that the soil can hold between these two limits or only that part of this water that is present at any given moment. It might be best to designate these two conditions by "plant-available-water capacity" and "plant-available water" respectively.

A frequently used term is "air-dry soil." This designates the moisture condition that approaches equilibrium with an atmosphere that is unsaturated with water. Consequently, depending upon the degree of saturation and the temperature, the water in an air-dry soil may be under a tension between pF 4.5 and 7, a truly wide margin. It is clear that the term "air dry" must be used with consideration of the given environment. It is not an equipotential point.

CLASSIFICATION OF SOIL MOISTURE

While it is recognized that there are no different kinds of soil water but that the water is merely held under different amounts of tension, it is practical to classify soil water into the conditions in which it exists.

In order to make it easier to understand the relationships between the various dimensions expressing soil-moisture tension and the soil-moisture constants, these items are presented together in Table 2-19.

Table 2-18 Classification of soil moisture

Water of constitution and interlayer water	Above pF 7
Hygroscopic water	pF 7–4.5
Capillary water	pF 4.5–2.5
Gravitational water	pF 2.5–0
Groundwater	Tension free

MOVEMENT OF SOIL WATER

THE DIFFERENT FORMS OF WATER MOVEMENT

Water can move in the liquid, gaseous, and solid phases:

1. *Liquid water*
   a. *Saturated flow:* Most pores are filled with water; this occurs in the zone of the groundwater and sometimes in the soil after heavy rains or during irrigation. Water in this condition is tension free.
   b. *Unsaturated flow:* Pores are partially filled with air. The water is under tension.
2. *Water vapor*
   a. *Diffusion:* Water vapor may move by diffusion as a result of vapor pressure (partial pressure) differences.
   b. *Mass flow:* Water vapor may flow in a mass with the other gases of the system in response to differences in total pressure.
3. *Ice:* Under ordinary conditions water movement in the soil does not occur in the form of ice. But formation of ice lenses makes the soil heave. Such movement of water in the solid phase only occurs as part of a movement of the entire soil body. This will not be included in the discussion of soil-water movement.

THE CONCEPTS OF FLOW

Water movement through soil is proportional to the product of the driving force and the conductivity of the soil for water. This movement, both liquid and vapor, can be expressed by the equation

$$Q = cDK$$

where Q = flow velocity
c = proportionality factor
D = driving force
K = conductivity of the medium

This relation holds true for heat transfer and for flow of electricity as well as for water movement.

THE DRIVING FORCE

The driving force in the case of water is a pressure gradient. Water moves from a position of high pressure to a position of low pressure. This is true for saturated and unsaturated liquid flow and for vapor flow.

In the case of saturated flow the pressure gradient may be brought about by differences in hydrostatic head (gravity, "water seeks its own level"). This pressure gradient may also be brought about by mechanical force (pressure from weight on soil surface or swelling colloids).

Table 2-19 Soil-moisture tension

Appearance of soil	pF	Tension equivalent to: cm of water	mbar	atm (approx)	ergs/g	Relative humidity 25 °C, %	Freezing point, °C	Soil pores: Equivalent diameter, mm	Size designation	Dominant function	pF	Soil moisture constants	Remarks
	7	10,000,000	9,800,000	10,000	98,000 X 10^5	0					7	Oven dry	
	6.5					10					6.5		
Dry	6	1,000,000	980,000	1,000	9,800 X 10^5	50			Hygros-copic surfaces	Unavailable soil moisture	6		
	5	100,000	98,000	100	980 X 10^5	93					5		
							−4.0						
	4.5	31,623				98		Colloidal size			4.5	Hygroscopic coefficient	Barely moist color
							−2.0						
	4.18	15,340	15,000	14.8		99	−1.22	0.0002			4.18	15 bar percentage; Wilting point	
	4	10,000	9,800	10	98 X 10^5						4		

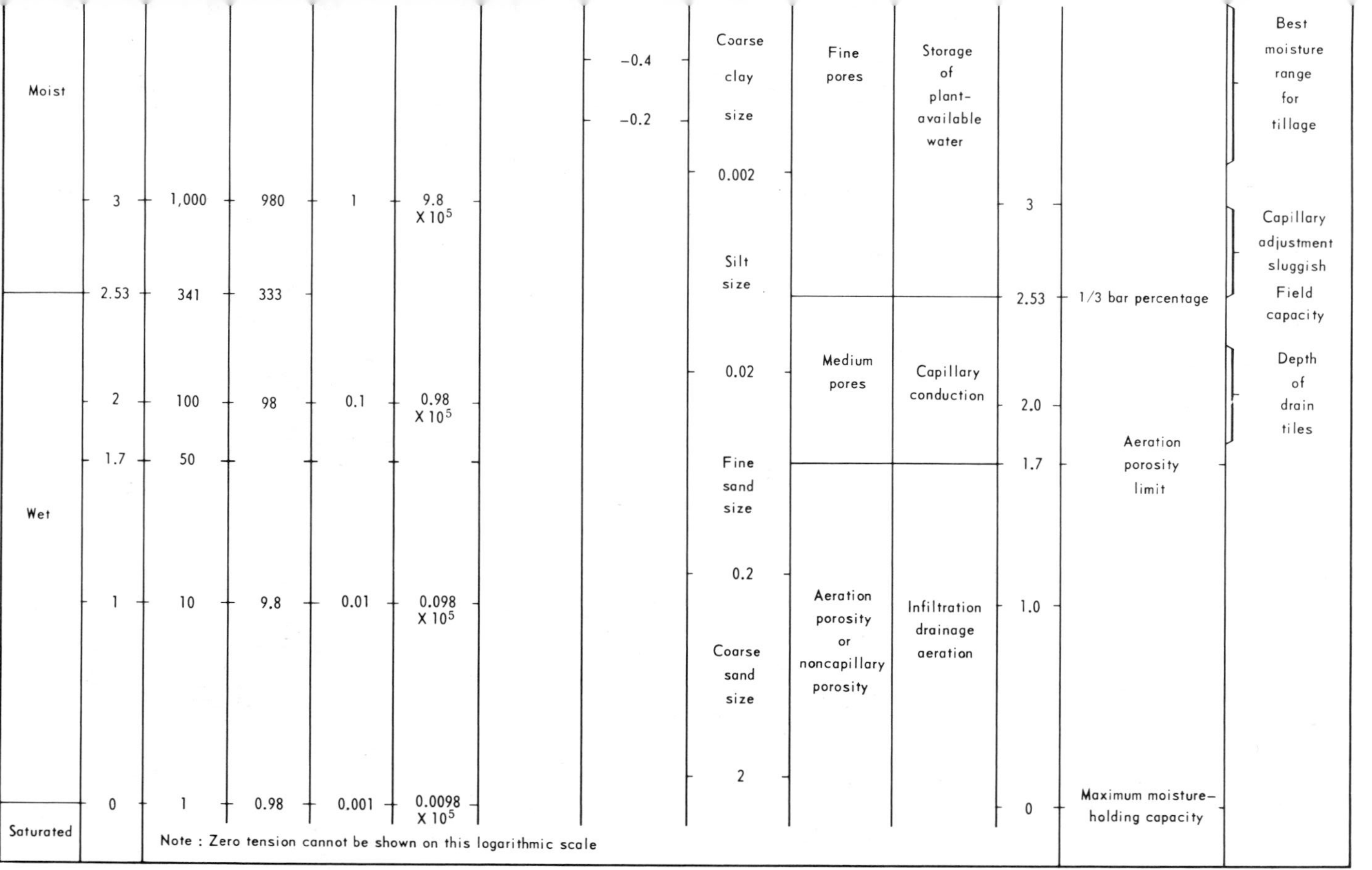
Moist
Wet
Saturated
3
2.53
2
1.7
1
0
1,000
341
100
50
10
1
980
333
98
9.8
0.98
1
0.1
0.01
0.001
9.8 X 10^5
0.98 X 10^5
0.098 X 10^5
0.0098 X 10^5
Note : Zero tension cannot be shown on this logarithmic scale
−0.4
−0.2
Coarse clay size
0.002
Silt size
0.02
Fine sand size
0.2
Coarse sand size
2
Fine pores
Medium pores
Aeration porosity or noncapillary porosity
Storage of plant-available water
Capillary conduction
Infiltration drainage aeration
3
2.53
2.0
1.7
1.0
0
1/3 bar percentage
Aeration porosity limit
Maximum moisture–holding capacity
Best moisture range for tillage
Capillary adjustment sluggish
Field capacity
Depth of drain tiles

In the case of unsaturated flow the pressure gradient is the sum of the difference in hydrostatic head and the difference in soil-moisture tension.

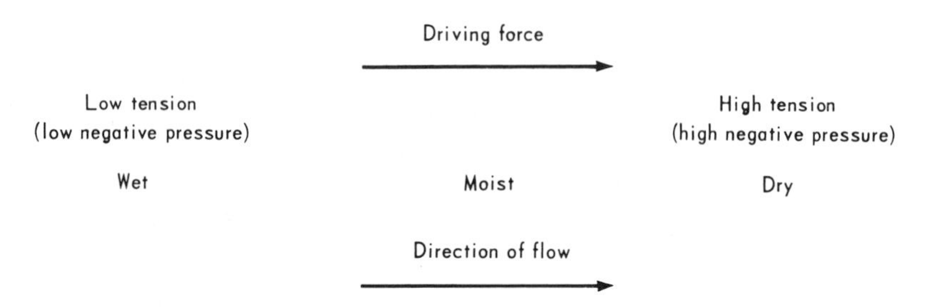

THE HYDRAULIC CONDUCTIVITY

The conductivity of soil for liquid water depends on the cross-sectional area of the pores and on the size of the pores. In saturated flow the conductivity increases as the fourth power of the radius. In unsaturated flow the conductivity depends on the degree of unsaturation. The drier the soil, the smaller is its conductivity.

We can conclude from these statements that the hydraulic conductivity is no simple function of porosity. Although the conductivity of a very porous soil is generally higher than that of a less porous soil as far as saturated flow is concerned, this relationship may be reversed for unsaturated flow.

The permeability of soil for water vapor is directly proportional to the volume of water-free pore space, regardless of size of pores. The reason for this is that the "mean free path" of water molecules is small compared to even the smallest pores that can be without water under normal conditions.

SATURATED FLOW

Saturated flow in soil occurs when the water is not under tension. All or most of the pores are completely filled with water. The rate of saturated flow, such as groundwater flow, is stated in two equations.

The law of Poiseuille expresses the flow of water in a narrow tube:

$$Q = \frac{P\pi R^4}{8LZ}$$

where Q = volume of flow, cc/sec
P = pressure difference, dynes/cm^2
R = radius of tube, cm
L = length of tube, cm
Z = viscosity of liquid, dyne-sec/cm^2 (poises)

Expressing this in words: The rate of flow of a liquid through a narrow tube is proportional to the fourth power of the radius of the tube and to the pressure and inversely proportional to the viscosity of the liquid and the length of the tube. Halving the diameter of the tube reduces the flow to one-sixteenth of its former rate. The Poiseuille equation is based on the assumption that the fluid in contact with the wall of the tube is at rest and that there is no turbulent flow.

The equation shows that the pore size is of outstanding importance, as its fourth power is proportional to the rate of saturated flow. This indicates that saturated flow under otherwise identical conditions decreases as the pore size decreases. In fact flow through deflocculated clay is practically zero. Generally the rate of saturated flow in soils of various textures is in this sequence:

Sand > loam > clay

The occurrence of viscosity in the Poiseuille equation indicates the effect temperature has on flow. Viscosity of water increases more than 1 percent with each degree-centigrade drop in temperature.

The law of Darcy states that the velocity of flow of a liquid through a porous medium is proportional to the force causing the flow and to the hydraulic conductivity of the medium. This can be expressed in several ways. The equation can be based on the pressure gradient as the driving force:

$$Q = \frac{cKAP}{L}$$

where Q = flow velocity (L^3T^{-1})
c = dimensionless proportionality constant
K = hydraulic conductivity ($M^{-1}L^3T$)
A = cross-sectional area (L^2) of the porous medium
P = pressure gradient ($ML^{-1}T^{-2}$)
L = length of the porous medium (L)

Or the equation can be based on the hydraulic-head gradient as the driving force:

$$V = Ki$$

where V = flow rate (LT^{-1})
K = hydraulic conductivity (LT^{-1})
i = hydraulic-head gradient (a dimensionless ratio)

It is noted that the dimensions of the hydraulic conductivity depend on the form of the equation. The actual magnitude of the hydraulic gradient or the coefficient of permeability has to be determined for every case. It depends not only on the nature of the porous medium but also

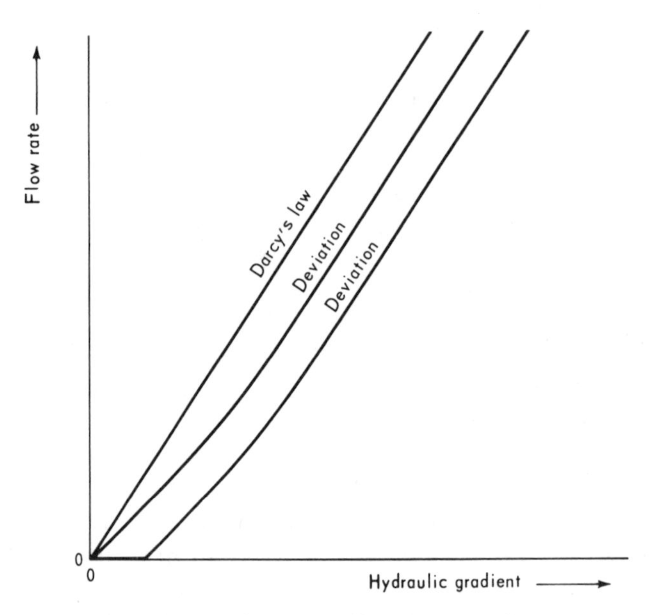

Fig. 2-12 Darcy's law of saturated flow through soil.

on the viscosity of the liquid. In the case of water, therefore, it depends on the temperature.

Darcy's law assumes a constant ratio hydraulic gradient/flow rate for saturated flow through porous media. Recent studies (Swartzendruber, 1962) indicate that in some cases low hydraulic gradients cause no or unproportionately small flow rates. This is especially true in clays (Miller and Low, 1963). It has been postulated that the reason for this may be that higher pressures are required to open larger passageways through the soil or that water near surfaces has an elevated viscosity.

Considering both the Poiseuille equation and Darcy's law we recognize that the conductivity for water does not depend upon the total amount of pore space. For example, a clay is a porous material but its conductivity is so low that it makes one of the best confining materials. The conductivity of sand and gravel depends much upon the sorting and grading of the particles composing the materials, those having grains of uniform size being the most permeable.

Saturated flow is of importance in drainage studies, but the rate of flow of water into a tile is greatly influenced by the rate of unsaturated flow, as the replenishment of the groundwater between the tiles depends on the rate of unsaturated downward movement of soil water. Since unsaturated-flow rates are much smaller than saturated-flow rates, tiles carry water a long time after the last rain.

UNSATURATED FLOW

The rate of unsaturated flow depends on the size of pores that are active and on the tension with which the water is held. While in saturated flow no air-water interface exists, meniscus formation and change has much to do with the rate of unsaturated flow. The tighter the water is held by the soil, the slower is it able to move under the influence of a tension gradient.

The less cross-sectional area of effective conducting pores is functioning, the slower is the movement. As the soil gets drier, the films around the particles become discontinuous and liquid flow stops completely.

Comparing a fine-textured and a coarse-textured soil, Gardner (1956) found that in the wet range up to about 1- or 1.5-bar tension (corresponding to pF 3 or 3.2) the sandy loam had greater hydraulic conductivity than the silty clay loam. In the moist range—when the soil-moisture tension was greater than this amount—this relationship was reversed. His data also show that this same tension represents a rather sharp boundary between rapid hydraulic conductivity in the wet range and very slow hydraulic conductivity in the moist range (Fig. 2-13).

Movement of unsaturated flow ceases in sand at a lower tension than in finer textured soils as the water films lose continuity sooner between the larger particles. The wetter the soil, the greater is the conductivity for water.

In the moist range unsaturated conductivity is greatest in fine-textured soils.

Sand < loam < clay

This is the reverse of the order encountered with saturated flow. However, in the wet range hydraulic conductivity occurs in the same or similar order as saturated conductivity.

Percolation ceases relatively quickly from a sandy soil, whereas it continues for a long time once a fine-textured soil profile has been wetted. The fact that springs flow throughout long periods of drought is proof that there is no sharp and definite tension value at which percolation ceases. In reality the rate of percolation decreases with the increase of tension with which the water is held and consequently with the decrease of the amount of water left in the soil. Percolation may be saturated flow, but more frequently it is unsaturated flow.

It is one of the great miracles of Nature that water is released quickly from the soil when the soil is wet and that the rate of removal is increasingly slower as the soil reaches the optimum moisture range for plants. This is a way to keep the moisture content of the soil near optimum con-

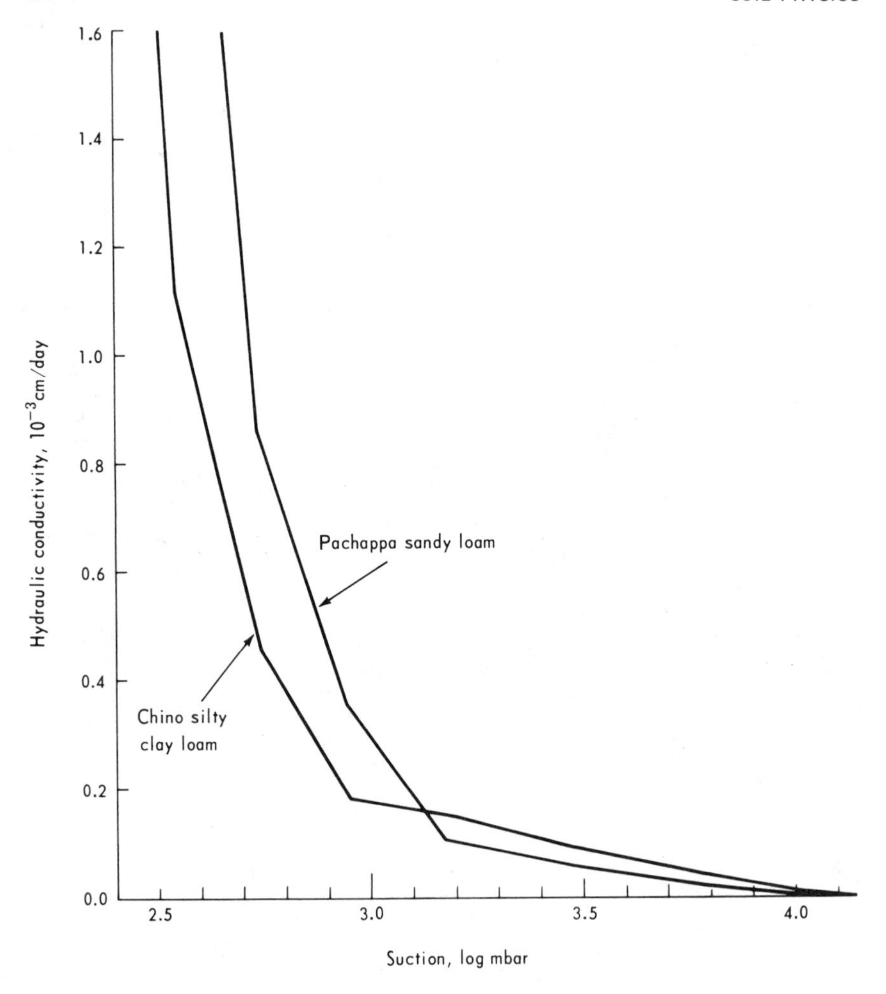

Fig. 2-13 Unsaturated hydraulic conductivity in moist and wet soil: In the wet range hydraulic conductivity is largest in the coarse-textured soil. In the moist range hydraulic conductivity is greater in the fine-textured soil. In this case the crossover point is near 1.3 bars. At the wilting point the hydraulic conductivity is practically zero in both soils. (*After Gardner,* 1956.)

ditions for a long time, to sustain stream flow, and to protect the valleys from floods.

VAPOR MOVEMENT

Two main forms of water movement in the vapor phase are

1. Bodily movement of soil atmosphere and
2. "Diffusion" or "distillation" of water vapor.

Bodily movement of the soil atmosphere can be caused by

Change in atmospheric pressure,
Change in soil temperature, and
Pressure change caused by infiltration (compression) and by percolation (evacuation).

Such mass flow of water vapor represents only a small portion of the entire water vapor movement in the soil. The extent to which it can occur is discussed in the chapter on soil air.

Diffusion of water vapor is caused by a vapor pressure gradient as the driving force. The vapor pressure of soil moisture is affected by

Moisture content: vapor pressure increases with moisture content,
Temperature: vapor pressure increases with temperature, and
Soluble salts: vapor pressure decreases with the increase of soluble salt content.

In "dry" soils, at the hygroscopic coefficient or drier, variations of moisture content are the outstanding causes for vapor pressure changes. Temperature differences are of secondary importance.

In "moist" or "wet" soils, at the wilting point or wetter, the relative humidity is so nearly 100 percent that changes in moisture content do not cause pronounced effects on vapor pressure. In this range vapor pressure changes are mostly brought about by temperature changes. But water vapor movement is pronounced only in the moist range. In the wet range vapor movement is negligible because there are few continuous open pores. In the dry range water movement exists, but there is so little water in the soil that the rate of movement is very small.

Water vapor movement goes on within the soil and also between soil and atmosphere. Examples are internal distillation, evaporation, condensation, and adsorption.

The rate of diffusion of water vapor through the soil is proportional to the square of the effective porosity, regardless of pore sizes.

The finer the soil pores, the higher is the moisture tension under which maximum water vapor movement occurs in soils. For instance, in the case of uniformly grained medium sand this occurs at approximately pF 2, while in a silt loam it occurs at approximately pF 3.5 (Jones and Kohnke, 1952). The reason for this difference is that two opposing factors are at play, the moisture content and the volume of water-free pores.

In a coarse-textured soil pores become free of liquid water at relatively low tensions. As such a soil dries out, only little water is available for vapor transfer. A fine-textured soil, especially if it is not well aggregated, has to reach fairly high tensions before a sufficient number of pores

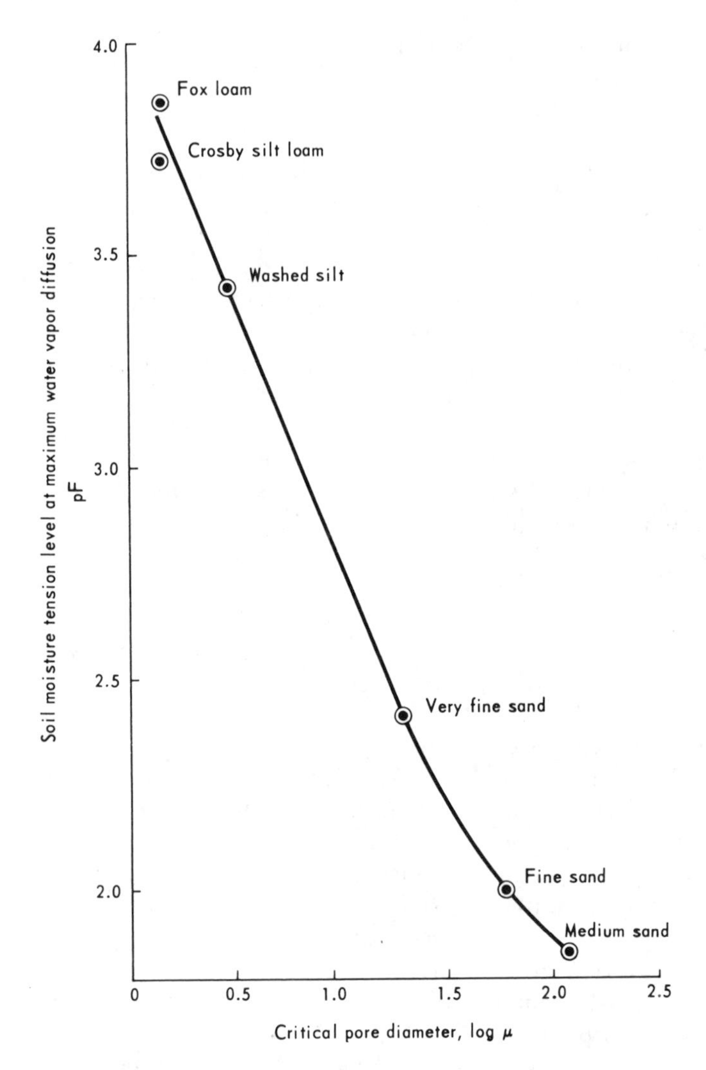

Fig. 2-14 Relation between pore size and water vapor diffusion: The maximum rate of water vapor diffusion in soils occurs when the pores of the size that represent the major portion of the pore space have been emptied of water, and the smaller pores still contain water. This is the "critical pore diameter." (*After Jones and Kohnke*, 1952.)

are freed from water. But it contains enough water even to the wilting point to permit vapor movement under the influence of a vapor pressure gradient.

It is interesting that maximum water vapor movement in soils occurs just before the wilting point is reached. This is the condition when such movement is of greatest importance for the growth and the survival of plants.

A very important phenomenon of water vapor movement is caused by differences in electrolyte concentrations. Salts reduce the vapor pressure of water. Hence, if they are localized in the soil, they create a vapor pressure gradient. Therefore water will have the tendency to move toward the salt. This tendency will be greater, the lower is the aqueous vapor pressure of the saturated solution of the salt in question. The vapor pressure gradients resulting from application of fertilizer cause water to move toward it, forming a moist layer around the fertilizer into which some of the dissolved salts can diffuse. In this zone plant roots find nutrients in concentrations that are favorable for uptake.

EFFECT OF SOIL MOISTURE ON PLANT GROWTH

TENSION

Water makes up four-fifths of the weight of a green plant. Of the rest, by far the greatest part is made up of compounds that are synthesized of water and carbon dioxide.

It is obvious that plants cannot grow in dry soil, but as the roots need air as well as water, mesophytic plants cannot grow in a wet soil for any length of time. Under conditions of adequate oxygen supply plant growth steadily decreases with the increase of soil-moisture tension from near saturation to the wilting point (Wadleigh, 1955; Danielson and Russell, 1957; Ginrich and Russell, 1957; Peters, 1957; and Giskin, 1965). The data represented in Fig. 2-15 show that the closer the moisture content is to saturation, the better will be plant growth as long as the roots receive sufficient oxygen. Since the relation between gaseous diffusivity and soil-moisture tension depends on the structure of the soil, it must be concluded that the optimum soil-moisture tension for plant growth will be lower on "porous" soils than on compact soils.

The ease of entry of water into the roots depends on the free-energy gradient from soil water to the water of the plant sap. It also depends on the transmissibility of the soil water.

The free energy or the total stress of soil water is determined by the matric suction (the attraction of the soil particles for the water) and by the osmotic components present. If the total stress is made up largely of matric suction, the transmissibility of water decreases rapidly as the

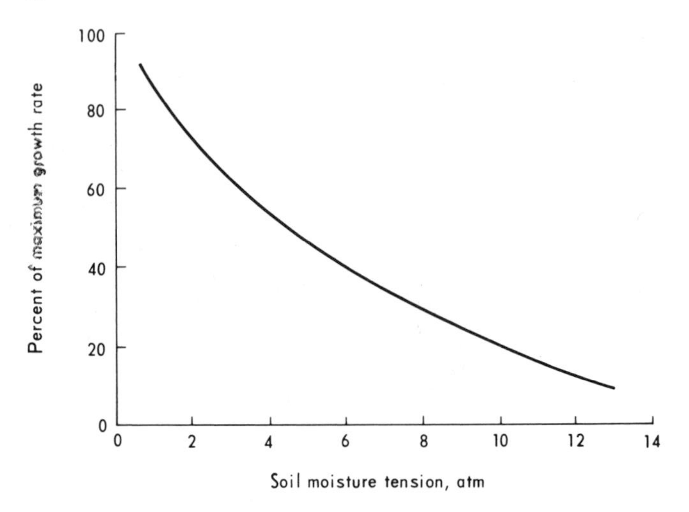

Fig. 2-15 The effect of soil-moisture tension on plant growth under conditions of adequate oxygen supply (average of several published data).

total stress increases. The reason for this has been explained in the preceding pages: Unsaturated hydraulic conductivity decreases greatly with decreasing water content. Where osmotic components are responsible for a large part of the total stress, transmissibility of water may continue at a high level even at relatively high water stresses. This means that roots have more rapid access to soil water and therefore plant growth is reduced much less than where the same stress is brought about largely by matric suction. If plants take up water in greater proportion than the electrolytes dissolved in the water, there will be an accumulation of electrolytes around the roots. This causes an increase in osmotic suction which reduces the free energy of the water and hence the availability of water to the roots. On the other hand, the increased osmotic suction brings about a tendency of the soil water to move toward the roots. Such factors as soil texture and structure, stage of growth, and temperature and relative humidity of the air have a great influence upon the best soil-moisture tension for any given plant species. It is an open question whether constant or variable soil-moisture tension is more favorable for plant growth.

TEMPERATURE

Temperature has a profound effect on the water-supplying power of soil. For the same moisture content the water-supplying power of soil is increased as the temperature is increased. The free energy of water increases with temperature. When moist soil is very cold, plants may suffer from physiologic drought if the atmosphere is warm and thus stimulates transpiration. When the soil is frozen, this condition is still more serious. Temperature also affects the activity of the microbes and hence their use of water.

FORCES INVOLVED IN WATER DELIVERY TO PLANTS

Imbibitional and osmotic forces bring about a powerful attraction of plant roots for soil water. It seems difficult to clearly separate the effects of these two types of forces, and it would be helpful to use one and the same system to express the energy levels of water attraction in the various parts of the plants and the energy levels of soil water. Transpiration is the original cause for building up these forces in the plants. It supplies the energy to pull water from the ground to great heights in the trees. That water can be lifted higher than 1,030 cm (1 atmosphere) is due to the great tensile strength of water that maintains a continuous column of water in the capillary conductive system of the plants.

ADAPTATION OF PLANTS TO SOIL-MOISTURE CONDITIONS

Plants have several ways of adjusting themselves to conditions of restricted water supply. Most crop plants when growing in rather dry land will not only have a smaller total weight, but also a smaller top/root quotient, than normally grown plants. On the other hand, plants growing in soil of high moisture content will have a large top/root quotient. Other adjustments are not of as obviously agronomic importance.

The type of root systems of the various plant species is frequently an indication of their adaptation to moisture conditions. The grasses of the dry prairies send their roots only to limited depths since soil moisture does not penetrate deeply. Desert plants have very shallow roots, enabling them to absorb quickly the water of sudden rainfalls or of condensation water formed at the soil surface. Alfalfa and sweet clover which originated in semiarid regions of Asia have very deep roots to forage throughout a large volume of soil. Many tree species also have deep root systems for the same reason. In wet soils the penetration of roots into deeper layers is frequently restricted by lack of air. Plants growing in swamps and poorly drained soils are shallow-rooted. Plants growing in dry and saline conditions (xerophytes and halophytes) develop special mechanisms to make them adapted to such extreme environment.

Crops that can grow successfully when the matric suction is high are called *drought tolerant.* Plants that can adjust to high solute suction are called *salt tolerant.* Aquatic plants are able to supply their roots with oxygen from their leaves. In this way rice can oxidize ferrous iron in the soil solution and precipitate it in the ferric form on the surface of the roots and thus prevent an excess of toxic material from entering the plant.

SOIL MOISTURE AND PLANT NUTRITION

Nutrient intake by plants increases with the increase of soil-moisture content from the wilting point to near saturation as long as no other growth factor becomes limiting.

The recognition of soil moisture–plant growth interrelationships is indispensable in fertilizer logistics: the art of having the right plant nutrients at the right spot at the right time. Plowing under fertilizer is an example of placing the fertilizer into a position where the moisture content will presumably remain greater than the wilting point throughout the year. Ample fertility decreases the water requirement of a crop. In the same way, ample moisture supply allows the plants to make good use of the available plant nutrients. Top dressing soil with fertilizer is generally of greatest benefit to meadow plants and weeds as these have much of their root systems in the topsoil.

Excess moisture by causing reduction in the soil slows down nitrification or inhibits it completely. It might even result in denitrification and loss of nitrogen in gaseous form. It also depresses the potassium intake of plants.

AMOUNT OF SOIL WATER AVAILABLE FOR PLANTS

That the amount of moisture available to plants determines the range of crop yield is quite apparent. In humid regions other growth factors frequently limit plant growth, but in the drier regions water is usually the first limiting factor. In the semiarid Northwest of the United States the wheat yield can fairly be predicted from the depth to which moisture has penetrated into the soil. The lower limit of annual precipitation needed to raise wheat every year in that area is about 350 mm. Where the rain of one year is insufficient to provide for a crop, the system of summer fallow is used to accumulate the water of two years for one crop. Even in many humid areas drought periods occur that may seriously affect crop growth.

The amount of soil moisture that is available to plants during a given period depends on

The depth and volume of soil penetrated by the roots,
The initial moisture content of each horizon,
The wilting point of each horizon, and
The replenishment of the soil water.

The depth of root penetration is determined by the nature of the plant, the mechanical penetrability of the soil, the water supply, the oxygen supply, and the nutrient supply. Generally speaking, roots penetrate deeply in porous, well-aerated soils of low moisture content.

The volume of soil available to each plant depends on plant spacing (population density) as well as on root penetration.

The initial moisture content of each horizon varies, of course, with the moment that is used as the beginning of observation.

Field capacity and wilting point determine the maximum amount of available water in the soil. These amounts vary from 5 percent of the total volume for gravelly soils and 10 percent for sandy soils to 15 and 20 percent for loams, silt loams, and clay loams.

The replenishment may occur as precipitation, condensation, adsorption, irrigation, or capillary rise from lower horizons.

At the time the crops are planted the moisture content at the surface must be considerably drier than field capacity, at least about pF 3.0 to 3.2, because otherwise they cannot be tilled and planted. Farther down, well-drained soils in the humid zone will be at field capacity, but soils with impeded drainage will be saturated or nearly so at some lower depth. The water table recedes during the vegetation period as a consequence of transpiration and possibly because of percolation. This means that soils with impeded drainage—assuming identical texture, structure, and root penetration—have more available water than well-drained soils.

EFFECT OF SOIL MOISTURE ON SOIL CONDITIONS AND SOIL DEVELOPMENT

SOIL DEVELOPMENT

Few if any components of soil-forming materials are entirely insoluble in water. When carbon dioxide is dissolved in water, as it usually is in the soil, the resultant carbonic acid makes the water a more efficient solvent. Hydrolysis and hydration assist in the breakdown of the parent material and of the soil itself. The products of decomposition are removed from the soil by the percolating water, or they may be redeposited at certain depths if the amount of water is not sufficient to keep them in solution or if pH changes induce precipitation.

Of the exchangeable cations sodium is leached out most readily. This is followed by magnesium and potassium, while calcium is held more tenaciously by the clay. The hydrogen ions in the percolating water take the place of these ions. The actual amounts of these ions lost from the soil by leaching—under Indiana conditions—are about 7 kg of potassium, 42 kg of magnesium, and 50 kg of calcium per hectare per year. That these quantities are not in the sequence of their ease of release is due to their vastly different amounts existing in exchangeable form in the soil (Weaver, Bushnell, and Scarseth, 1948).

Iron and aluminum are leached out of the surface layers of the soil if the percolate is acid in reaction. In the case of iron, removal by leaching is particularly large where reduced conditions exist in the soil. If the percolate is alkaline—due to rapid oxidation of organic matter—silicon is removed and iron and aluminum stay behind. The higher solubility of carbonic acid in cold water than in warm water may explain the acid leaching of the podzols and podzolic soils and the nearly neutral leaching of the latosols. Silica is soluble in warm water, if it is alkaline or neutral in reaction, while it is essentially insoluble in cold acid water as it occurs in the period of strongest leaching (the spring) in the typical podzolic regions. In the case of the leaching of iron and aluminum, the effect of the reaction of the water is exactly the opposite of what it is in the case of silica.

If leaching continues for very long periods of time and the surface soil is not removed by erosion (geologic erosion), deep layers of the most insoluble and hence infertile components of the parent material form the topsoil. Very old soils are therefore generally not fertile, while also very young soils in which not much clay and not much organic matter have yet formed are not very productive. The best soils are usually those of intermediate age. An exception constitutes most of the recent alluvial soils which, though very young, are really redeposited soils and generally very fertile.

Climate has an outstanding effect upon the type of soils formed. Important especially are moisture and temperature. The quantity of moisture available for soil formation depends both on the amount and distribution of annual precipitation and on the temperature. The higher the temperature, the more energy exists to evaporate the water and the less water is available for the soil-forming process. It has been pointed out previously (page 37) that a modification of the Lang rainfall factor, when plotted against average annual temperature, is useful to classify the influence of climate upon soil development. Figure 2-16 indicates under what climatic conditions some of the major soil groups occur.

Marbut (1927) recognized the major importance of water on soil formation, when he grouped the soils that are not leached free of carbonate as pedocals and those that are as pedalfers. The soils in the humid and perhumid climates in Fig. 2-16 are pedalfers, the others pedocals. "Pedalfer" and "pedocal" are terms no longer used by the Soil Survey Staff of the U.S. Department of Agriculture (1960).

From the point of view of structure and fertility, the groups of intermediate moisture conditions, i.e., the Black earths and the Prairie soils are the best. They are not as thoroughly leached of nutrients as the podzolic and latosolic groups, yet they have developed under sufficient influence of moisture that they contain much organic matter and sufficient clay so that the crops do not suffer from drought or accumulation of electrolytes as do the Desert, the Brown, and the Dark-brown soils.

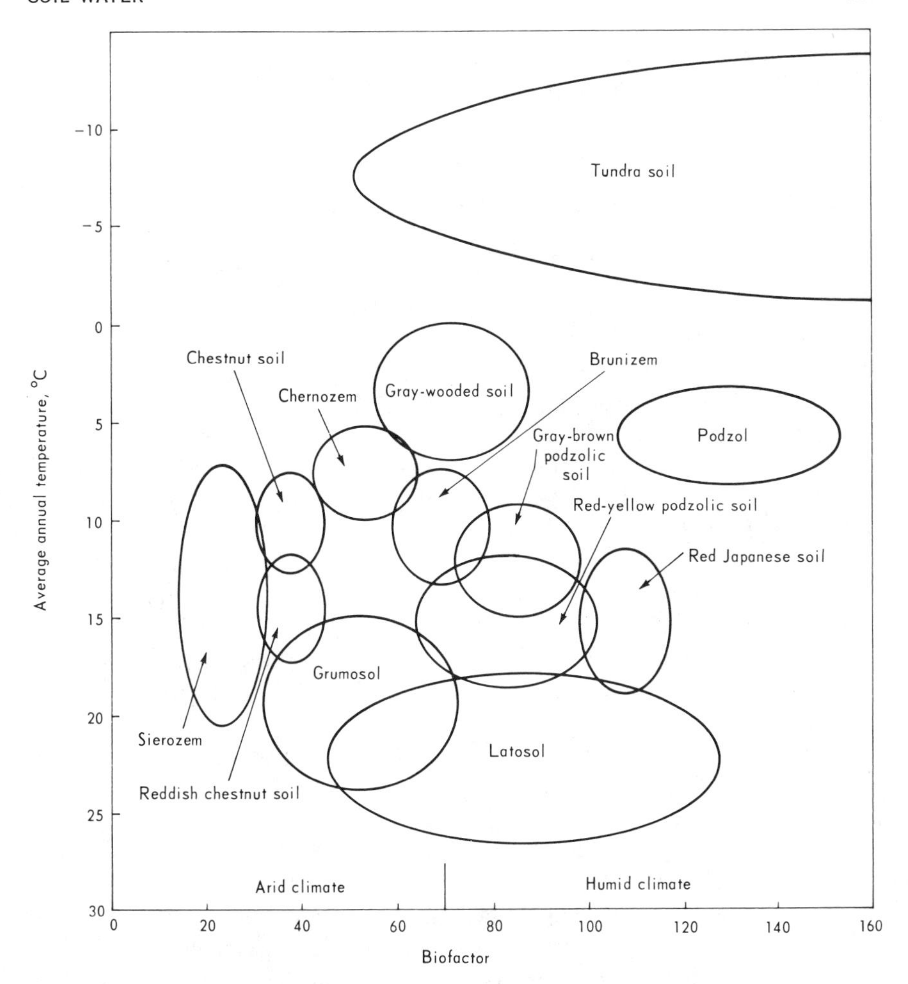

Fig. 2-16 Climate-soil relations: Each area describing the climatic conditions under which a given Great Soil Group occurs is delineated by the standard deviations from the means of temperature and biofactor for all samples of that group. (*After H. Kohnke, R. G. Stuff, and P. A. Miller, Quantitative Relations between Climate and Soil Formation. "Zeitschrift f. Pflanzenernährung und Bodenkunde," vol.* 119, *pp.* 24–33, 1968.

EFFECT OF TOPOGRAPHY ON SOIL WATER AND SOIL PROFILE

Within the Great Soil groups hydromorphic sequences of soil series occur as a consequence of parent material and topographic position. An example of the classification of soils of the same parent material according to the topographic effects is the system by Bushnell (1958). He recognizes 10 possible hydromorphic profiles in the Catenas of the Gray-brown podzolic soils and the Prairie soils of Indiana. The outstanding ones are these four:

1. *"II Profile"* Soils of the flat upland with slopes up to 2 percent. They have only little runoff under pristine conditions. The amount of infiltration plus evapotranspiration is nearly the same as precipitation. Clay is washed into the *B* horizon forming a clay pan. Surface soils are gray; subsoils are acid because they are leached of bases and they are mottled due to restricted drainage.
2. *"IV Profile"* Soils of medium slopes, about 4 to 15 percent. Some water from the larger rainstorms is swept downhill, taking dissolved mineral material with it. Clay formation in the *B* horizon is not very pronounced. Internal drainage is good, therefore the surface soil color is gray brown and the subsoil rust brown and usually acid. As these soils keep less water than they receive in the rain, they tend to be droughty.
3. *"VI Profile"* Soils of the steep slopes, steeper than 20 percent. Runoff and geologic erosion are so pronounced that only little of the leached material of the surface soil enters the subsoil. Before any recognizable accumulation of clay in the subsoil can take place, the weathered surface soil is removed by geologic erosion and the newly exposed soil is leached out. In this way a soil profile with *A* and *C* horizons, but without a *B* horizon, is formed. Due to excessive runoff, the erosion hazard is great and these soils are adapted only to permanent vegetation.
4. *"VIII Profile"* Soils of the swales, 0 to 1 percent slopes. Not only the direct rainwater, but also the enriched seepage and runoff water from the surrounding upland helped to develop these soils. They were moist practically all the time, before drainage ditches were dug; abundant vegetation grew and resulted in a high content of organic matter. This humus mitigates the effect of the rather high clay content. These soils are black to about 30 cm down, then gray and yellow mottled. If well drained, they are very fertile. They are nearly neutral in reaction throughout the profile.

EFFECT OF WATER ON SOIL CONDITIONS

Aeration The pore space of any soil is filled partially with air, partially with water. The space taken up by water cannot be filled with air. The amount of air in turn affects the composition of the soil air and the oxidation-reduction conditions of the soil. If water fills most of the pores of the soil and no oxygen can enter, the soil becomes *waterlogged*. This condition is particularly serious in the warm season, when roots and microbes are very active and require much oxygen. Nonaerobic microbes cause reduction of the soil by using oxygen from compounds and bring about chemical changes that are detrimental for soil and plants. Ferric iron changes to ferrous iron, manganese compounds become manganous compounds. Methane, hydrogen, hydrogen sulfide, and methyl sulfide are formed. Most of these are toxic to plants. The organic glues that have been produced by aerobic microbes and that serve to protect soil aggregates are decomposed and no new ones can be formed. The soil structure therefore deteriorates. Pore spaces are clogged and water and air regimes suffer even after the soil has dried up.

Temperature As the specific heat of water is about five times as high as that of the mineral soil particles, it takes much more energy to heat a wet soil than a dry soil. The amount of water in the soil has a pronounced effect on the influence of freezing and thawing on soil temperatures. Because of the energy transfer associated with the change of state of water the temperature will stay much more nearly the same in a soil full of water than in one that is only slightly moist. More on the effect of soil moisture on soil temperature is discussed in Chap. 8.

Hydrology The infiltration capacity of a wet soil is smaller than that of a moist soil. A rain will therefore cause runoff much easier from a wet soil than from an originally moist soil. On the other hand, a very dry soil, especially if it is high in organic matter, has a smaller infiltration capacity than when it is moist. This is due to *capillary depression* caused by coating of soil particles with water-repelling gums and resins.

Soil structure Very important is the lubricating action of wetted clay colloids on the changes of soil structure, e.g., puddling. The best soil structure results from intermediate moisture supply.

Tillability The moisture content of soil is of outstanding influence upon its tillability. Intermediate moisture conditions are best for this purpose. This is the range of friability of the soil.

The details on most of the items mentioned in these last paragraphs are discussed elsewhere in this text: viz., classification of soil moisture, soil structure, soil temperature, soil air, soil-water management.

METHODS OF MEASURING AND EXPRESSING SOIL MOISTURE

THE PURPOSES OF SOIL-MOISTURE DETERMINATIONS

Although there are many ways in which soil moisture can be expressed, there are fundamentally only two principles according to which this is done: the amount of water in a given amount of soil, and the stress or tension under which the water is held by the soil. The relationship between these two properties throughout the entire moisture range gives

Table 2-20 Purposes and types of soil-moisture determinations

Purpose of soil-moisture determination	*Required type of soil-moisture determination*
Plant-available water	The amounts of water at two limiting tensions
Irrigation	The tension at which irrigation should be started
Drainage and aeration	The tension to which it is desirable to drain the soil
Tillage	The tension range at which the soil benefits most from tillage
Thermal capacity of soil	The amount of moisture in the soil
Thermal conductivity of soil	The tension of soil moisture is closely related to the thermal conductivity of the soil
Soil-moisture research and determination of soil-water budget	The knowledge of both amounts and tensions is necessary

a good deal of insight into the physical characteristics of a soil. The relationship can be shown in a so-called soil-moisture-retention curve.

Whether it is more important to know the amount of water in the soil or the stress or tension under which it is held depends on the purpose for which the moisture determination is made. Table 2-20 shows that in the majority of cases a knowledge of the soil-moisture tension is required.

AMOUNTS OF WATER

The amount of water that is held by a certain mass or volume of soil is determined and can be expressed as weight percent or volume percent. Soil moisture in weight percent is based on the dry weight of the sample. The weight-percent determination is simpler, but the volume-percent determination gives a better concept of the moisture that the roots can reach.

Note: $$\text{Weight percent} = \frac{\text{weight of water} \times 100}{\text{weight of oven-dry soil}}$$ (using the same weight units)

$$\text{Volume percent} = \frac{\text{volume of water} \times 100}{\text{volume of soil sample}}$$

or

$$\text{Volume percent} = \text{volume weight} \times \text{weight percent}$$

Gravimetric method The standard method of determining the amount of water in soil is to place the weighed sample in an oven at 105°C and dry it to constant weight. The weight difference is considered to be water. An error can result from the oxidation of organic matter.

After the soil sample is removed from the oven, it is cooled to room temperature at extremely low water vapor pressure in a dessicator over calcium chloride or another drying agent. Under certain conditions where special accuracy is required soils can be dried in a vacuum or over drying agents such as phosphorus pentoxide.

Several methods have been developed to simplify the determination of the amounts of soil moisture. One of these is to mix the soil with a known amount of calcium carbide. Calcium carbide reacts with the water and removes it in the form of acetylene gas. The moisture is determined by the difference in weight. This method has been recommended for field work because of its speed, but it is not very satisfactory. More recently infrared radiation is utilized for drying soil samples quickly.

Neutron scattering Soil-moisture content can also be determined by the neutron-scattering technique (Stone *et al.*, 1955). Fast neutrons lose energy and are slowed down when they collide with hydrogen atoms. The slow neutrons are caught by a detector and counted by a counting unit. This method enables one to determine the moisture content of a soil with only a negligible effect on the moisture content. It can therefore be repeated indefinitely. The volume of soil represented in the measurement is rather large and therefore this technique does not lend itself to a study in which moisture changes in small areas are of importance.

Gamma ray attenuation Moisture changes in soil can be determined by gamma ray attenuation (van Bavel *et al.*, 1957). This technique depends on the fact that gamma rays loose part of their energy upon striking another substance. Gamma rays that are emitted from radioactive isotopes, e.g., $cesium^{137}$, can be focused into narrow beams, so that they can be used to study moisture in layers as thin as 1 cm. In contrast to this, neutron scattering measures moisture in a fairly large volume, a sphere of about 15-cm radius.

The disadvantage of the gamma ray attenuation technique is that gamma rays are loosing energy by hitting any substance, not just hydrogen as is the case with neutrons. For this reason gamma ray attenuation is useful particularly where only *changes* of moisture content are to be determined. If it is to be used to measure absolute moisture contents, each soil to be studied has to be calibrated.

SOIL-MOISTURE TENSION

Two approaches are used in studying soil-moisture tension. Either the tension level is determined in a soil as it is found or a specific tension is established in soil. In both cases the investigator determines the amount of water in the soil as well as its tension. A variety of methods exists that can be used for the determination of soil-moisture tension. These have been discussed previously (pages 46–49) and it has been pointed out to which tension range each of these methods applies.

ESTIMATING SOIL-MOISTURE CONDITIONS

It is not always practical to quantitatively determine soil moisture. In many cases an estimate is sufficient. This should, of course, come as near the actual situation as possible. Visual observation combined with simple field tests is useful. When the water in the soil glistens, the soil is wetter than field capacity. When it is possible to mold a ball of the soil, it has a tension of less than 1 atmosphere. When the soil crumbles, it has a tension of over 1 atmosphere. When the soil has a light color, it is drier than the hygroscopic point. At the wilting point the soil is crumbly, feels slightly moist, and has a dark color (though somewhat lighter than at 1 atmosphere tension).

Soil moisture can be estimated also by its effect on plants. Direct visual observation indicates only the extreme conditions of drowning and wilting. A more precise index of the soil-moisture situation can be obtained by the determination of the water status of the plants. One method calls for placing a drop of a mixture of xylol and mineral oil on the underside of a leaf and watching the entry of this liquid into the leaf. The two liquids are mixed in different proportions from pure xylol to pure mineral oil. When a mixture with a high proportion of mineral oil readily enters the leaf, this indicates that the stomata are open and that there is plenty of water available to the plant. A plant that suffers from lack of water closes its stomata and only the much less viscous xylol can enter. This technique is useful for some crops in the hands of an experienced observer.

Mederski (1962) measures the resistance of a plant leaf to beta ray penetration in order to draw conclusions upon the moisture situation of the soil. The more water is in the leaf, the smaller is the penetration.

REPORTING SOIL-MOISTURE DATA

The most valuable information concerning soil moisture is the relationship between moisture tension and moisture content. Wherever possible both of these parameters should be reported. Figure 2-17 gives an example of moisture-retention relationships in the form of moisture-release curves for three soils of different texture and structure.

Such desorption curves give ready information on the physical characteristics of soils. The large volume of aeration porosity and the low available-moisture-holding capacity show the Plainfield sand to

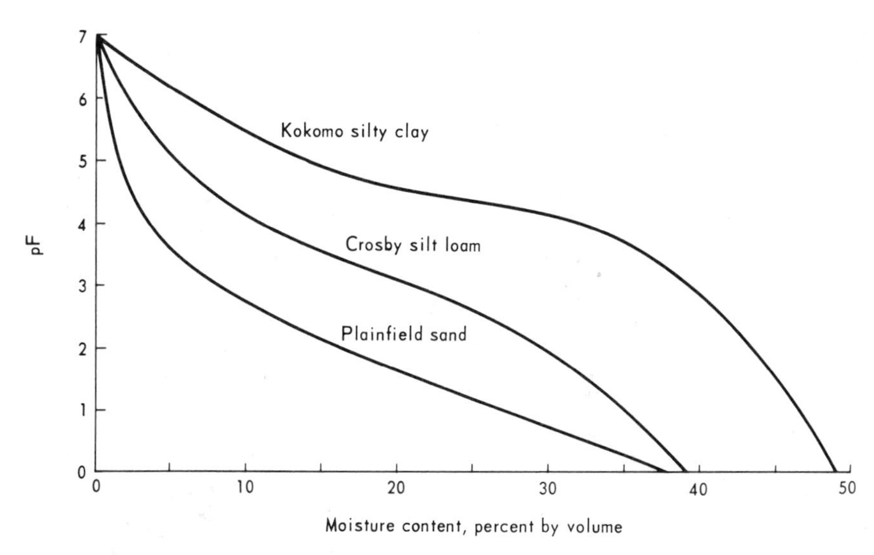

Fig. 2-17 Moisture-desorption curves for three soils.

have high infiltration capacity, but to be droughty. The Crosby silt loam has intermediate infiltration capacity and is well endowed with available-moisture-holding capacity. Its medium texture can be deduced from the amount of hygroscopic water. The curve for the Kokomo silty clay shows restricted aeration porosity, intermediate available-water-holding capacity, and a very high clay content (actually 46 percent clay). The organic matter content in both the Plainfield sand and the Kokomo silty clay is low. In the former this can be concluded from the low content of hygroscopic water, in the latter from the low aeration porosity.

Table 2-21 is presented to suggest guidelines for the interpretation of soil-moisture-retention data.

SOIL–WATER MANAGEMENT

The moisture situation of a soil can be modified in a variety of ways. The methods used may be direct application or removal of water or may affect the moisture content through changes of structure or temperature of the soil. The most important methods are discussed in Chap. 10. For soil physics to be helpful in soil-water management, close recognition must be given to the tension of the water. Many articles are still published in which only the amount of soil moisture is stated and no, or inadequate, reference is made to moisture tension.

Table 2-21 Guide for the interpretation of soil-moisture-retention data

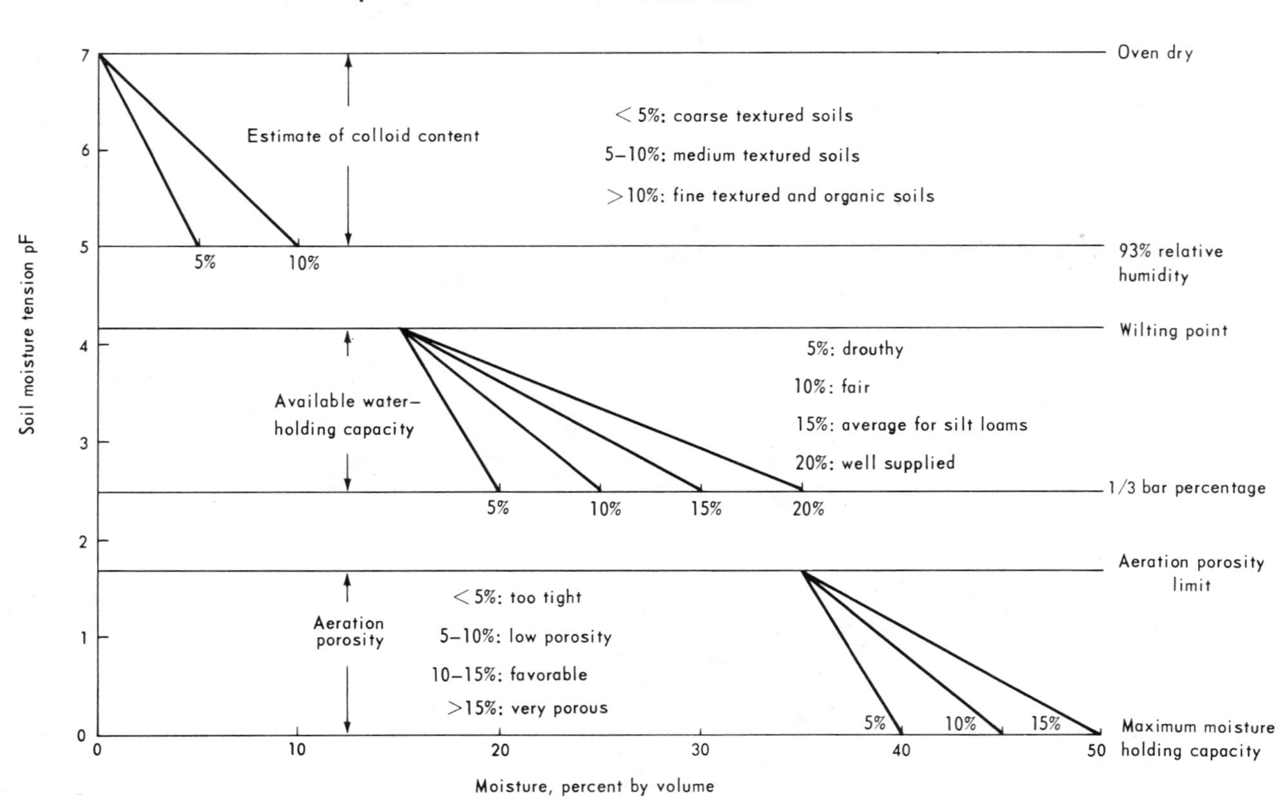

REFERENCES

Adam, N. K.: "The Physics and Chemistry of Surfaces," 3d ed., Oxford University Press, London, 1941.

Bouyoucos, G. J., and A. H. Mick: An Electrical Resistance Method for the Continuous Measurement of Moisture under Field Conditions, *Michigan State College Tech. Bull.* 172, 1940.

Brawand, H., and H. Kohnke: Microclimate and Water Vapor Exchange at the Soil Surface, *Soil Sci. Soc. Am. Proc.*, vol. 16, pp. 195–198, 1952.

Briggs, L. J., and J. W. McLane: The Moisture Equivalents of Soils, *U.S. Dept. Agr. Bur. Soils Bull.* 45, 1907.

——— and ———: Moisture Equivalent Determinations and Their Applications, *Proc. Am. Soc. Agron.*, vol. 2, pp. 138–147, 1910.

——— and H. L. Shantz: The Wilting Coefficient for Different Plants and Its Indirect Determination, *U.S. Dept. Agr. Bur. Plant Ind. Bull.* 230, 1912.

Buckingham, E.: Studies on the Movement of Soil Moisture, *U.S. Dept. Agr. Bur. Soils Bull.* 38, 1907.

Bushnell, T. M.: "A Story of Hoosier Soils," Peda-Products, West Lafayette, Ind., 1958.

Committee on Terminology of the Soil Science Society of America: Report of Definitions Approved by the Committee on Terminology, *Soil Sci. Soc. Am. Proc.*, vol. 20, pp. 430–440, 1956.

Danielson, R. E., and M. B. Russell: Ion Absorption by Corn Roots as Influenced by Moisture and Aeration, *Soil Sci. Soc. Am. Proc.*, vol. 21, pp. 3–6, 1957.

De la Hire, Philippe: Remarques sur l'eau de la pluie, sur l'origine des fontaines; avec quelques particularités sur la construction des citernes, *Hist. de l'Acad. Roy. des Sci. Ann.*, pp. 56–59, 1703.

Dorsey, N. E.: "Properties of Ordinary Water Substance," Am. Chem. Soc. Monograph Ser., Reinhold Publishing Corporation, New York, 1940.

Gardner, W. R.: Calculation of Capillary Conductivity from Pressure Plate Outflow Data, *Soil Sci. Soc. Am. Proc.*, vol. 20, pp. 317–320, 1956.

Ginrich, J. R., and M. B. Russell: A Comparison of Effects of Soil Moisture Tension and Osmotic Stress on Root Growth, *Soil Sci.*, vol. 84, pp. 185–194, 1957.

Giskin, M. L.: "Corn Root Response to Soil Suction," unpublished Ph.D. thesis, Purdue University, Lafayette, Ind., 1965.

Holdridge, L. R.: Determination of Atmospheric Water Movements, *Ecology*, vol. 43, pp. 1–9, 1962.

Jones, H. E., and H. Kohnke: The Influence of Soil Moisture Tension on Vapor Movement of Soil Water, *Soil Sci. Soc. Am. Proc.*, vol. 16, pp. 245–248, 1952.

Kohnke, H., and A. R. Bertrand: Rainfall Characteristics, "Soil Conservation," pp. 73–82, McGraw-Hill Book Company, New York, 1959.

Lang, R.: Versuch einer exakten Klassifikation der Böden in klimatischer und geologischer Hinsicht, *Intern. Mitt. Bodenkunde*, vol. 5, pp. 312–346, 1915.

Lebedev, A. F.: Relation of Water to Soil, *Second Intern. Congr. Soil Sci.*, vol. 6, pp. 65–88, 1930.

Low, P. F.: Physical Chemistry of Clay-Water Interaction, *Advan. in Agron.*, vol. 13, pp. 269–327, 1961.

Marbut, C. F.: A Scheme for Soil Classification, *First Intern. Congr. Soil Sci.*, vol. 4, pp. 1–31, 1927.

Mederski, H. J.: Determination of Internal Water Status of Plants by Beta Ray Gauging, *Soil Sci.*, vol. 92, pp. 143–146, 1961.

Miller, R. J., and P. F. Low: Threshold Gradient for Water Flow in Clay Systems, *Soil Sci. Soc. Am. Proc.*, vol. 27, pp. 605–609, 1963.

Peters, D. B.: Water Uptake of Corn Roots as Influenced by Soil Moisture Content and Soil Moisture Tension, *Soil Sci. Soc. Am. Proc.*, vol. 21, pp. 481–484, 1957.

Richards, L. A., and L. R. Weaver: Fifteen Atmosphere Percentage as Related to the Permanent Wilting Percentage, *Soil Sci.*, vol. 56, pp. 331–339, 1943.

Robertson, L. S., and H. Kohnke: The pF at the Wilting Point of Several Indiana Soils, *Soil Sci. Soc. Am. Proc.*, vol. 11, pp. 50–52, 1946.

Schofield, R. K.: The pF of the Water in Soil, *Third Intern. Congr. Soil Sci.*, vol. 2, pp. 37–48, 1935.

Soil Survey Staff: Soil Permeability, "Soil Survey Manual," *U.S. Dept. Agr. Handbook* no. 18, pp. 167–168, 1951.

———: "Soil Classification, a Comprehensive System," 7th Approximation, Soil Conservation Service, U.S. Department of Agriculture, 1960.

Stone, J. F., D. Kirkham, and A. H. Read: Soil Moisture Determination by a Portable Neutron Scattering Moisture Meter, *Soil Sci. Soc. Am. Proc.*, vol. 19, pp. 419–423, 1955.

Swartzendruber, D.: Non-Darcy Flow Behavior in Liquid-saturated Porous Media, *J. Geophys. Res.*, vol. 67, no. 13, pp. 5205–5213, 1962.

Van Bavel, C. H. M., N. Underwood, and S. R. Roger: Transmission of Gamma Radiation by Soils and Soil Densitometry, *Soil Sci. Soc. Am. Proc.*, vol. 21, pp. 588–591, 1957.

Wadleigh, C. H.: Soil Moisture in Relation to Plant Growth, *U.S. Dept. Agr. Yearbook of Agr.*, pp. 358–361, 1955.

Weaver, R. A., T. M. Bushnell, and G. D. Scarseth: Correlative Studies of Indiana Soils, I, Some Physical and Chemical Characteristics of Seven Major Soil Types, *Soil Sci. Soc. Am. Proc.*, vol. 13, pp. 484–493, 1948.

Wischmeier, W. H.: A Rainfall Erosion Index for a Universal Soil-loss Equation, *Soil Sci. Soc. Am. Proc.*, vol. 32, pp. 246–249, 1959.

——— and D. D. Smith: Predicting Rainfall-erosion Losses from Cropland East of the Rocky Mountains, *Soil Sci. Soc. Am. Proc.*, vol. 23, pp. 246–249, 1965.

3 MECHANICAL COMPOSITION OF THE SOIL

While soil is a three-phase system, *mechanical composition* refers only to its solid phase. This is made up of mineral and organic components. In this chapter the main stress is put on the mineral fraction. A special chapter is devoted to organic matter. "Mechanical composition" here refers to the ultimate particles, such as sand, silt, and clay, not to the aggregates. These are discussed in Chap. 5. Knowledge of the mechanical composition of a soil gives an insight into its physical, chemical, and biological potential. It is a vital aid in managing the soil for plant growth as well as for engineering purposes.

CLASSIFICATION OF SOIL PARTICLES

Mineral soil particles can be classified according to size, shape, density, or chemical composition.

SIZE

Generally material larger than 2 cm in diameter is called stones, material between 2 cm and 2 mm in diameter is called gravel, while the material smaller than 2 mm in diameter is the *fine earth*. Only this latter is normally considered in chemical and mechanical analyses of soils. The components of fine earth are sand, silt, and clay. The boundary between silt and clay has generally been set at 2 microns (0.002 mm) in diameter. The boundary between sand and silt is taken at either 0.02 or 0.05 mm in diameter. The size limits of the soil-particle fractions have been established by various national and international organizations.

The international classification, originally proposed by Atterberg, is:

Fraction	*Diameter, mm*
Coarse sand	2–0.2
Fine sand	0.2–0.02
Silt	0.02–0.002
Clay	< 0.002

The U.S. Department of Agriculture classification is:

Fraction	*Diameter, mm*
Very coarse sand	2–1
Coarse sand	1–0.5
Medium sand	0.5–0.25
Fine sand	0.25–0.1
Very fine sand	0.1–0.05
Silt	0.05–0.002
Clay	< 0.002

The clay is sometimes subdivided into coarse clay of 0.002 to 0.0002 mm in diameter and fine or colloidal clay of less than 0.0002 mm in diameter. The word "colloidal" refers to the size of matter. The colloidal size is intermediate between matter that is visible under an optical microscope and invisible molecules. These two classifications are shown in Table 3-1 on a logarithmic scale together with a modification of the international classification, recently used in Europe. It appears that the U.S. Department of Agriculture method gives overly much emphasis to the sand fraction while neglecting the silt fraction.

The original international classification has probably too few fractions. The more detailed classification seems preferable, as it gives practically the same size range to each fraction. Its main subdivisions—sand, silt, and clay—coincide essentially with those of the U.S. Department of Agriculture method. Its disadvantage is the amount of work required for such an analysis.

Table 3-1 Methods of soil-particle fractionation

Diameter, mm	USDA	*International*	*European*	*Diameter, mm*
2.0				2.0
	Very coarse sand		Coarse sand	
1				
	Coarse sand	Coarse sand		0.6
0.5				
	Medium sand		Medium sand	
0.25				
				0.2
	Fine sand		Fine sand	
0.1				
	Very fine sand	Fine sand		0.06
0.05				
			Coarse silt	
				0.02
	Silt		Medium silt	
		Silt		0.006
			Fine silt	
0.002				0.002
			Coarse clay	
				0.0006
	Clay	Clay	Medium clay	
				0.0002
			Fine clay	

SHAPE

Silt and sand particles generally approximate spherical or cubical shape. They are the minerals that were originally crystallized to form rocks. Weathering has separated them from their original compound. Their sharp edges are smoothed off by chemical decomposition and by physical influences, such as freezing and thawing or rubbing in a stream bed.

Transportation in a glacier scratches the surfaces of sand grains and gives them imprints characteristically different from the smooth surface of river sand. Soil that has formed in place from igneous rocks and has not been transported contains sand and silt particles with sharp edges.

Clay particles are plate or lath shaped depending on the crystal structure of the specific type of clay. They are never spherical or cubical. Some of the smallest soil particles are shapeless (amorphous).

According to Wadell (1932) the shape of sand grains can be quantitatively expressed in terms of "sphericity" and "roundness." "Sphericity" is the cube root of the ratio of the volume of the particles divided by the volume of the smallest sphere into which the particle would fit. The sphericity of a perfect sphere is 1. The sphericity of a thin needle-shaped particle approaches zero. "Roundness" expresses the angularity of the particle. It is measured as the ratio of the average radius of corners and edges of the particle divided by the radius of the maximum circle that can be inscribed in the particle. In a particle with sharp edges and corners the radii are small compared to that of the inscribed circle, and the ratio is small. If the edges and corners are smooth, their radii will be nearly the same as that of the inscribed circle and the ratio will approach 1. Both sphericity and roundness can be judged to a satisfactory degree of accuracy by comparing sand on a glass slide under a microscope with a chart showing examples of particles of varying sphericity and roundness. This system of classifying soil-particle shapes is distinctly superior to the use of qualitative descriptions such as round, smooth, angular, although in many instances these expressions will be sufficient.

DENSITY

The density of the dominant soil minerals such as quartz and several of the feldspars is near 2.65 g/cc. Consequently the *particle density* of many soils with a small percentage of organic matter is near 2.65 g/cc. Since the density of the soil organic matter is usually between 1.3 and 1.5 g/cc, the actual particle density of soil with a sizable humus content is determined by the densities and the relative proportion of both the mineral and the organic soil components.

This can be stated as

$$d_s = d_m p_m + d_o p_o$$

where d_s = density of the soil
d_m = density of the mineral components
d_o = density of the organic components
p_m = fraction of the mineral components in the soil
p_o = fraction of the organic components in the soil

Table 3-2 Densities of several soil constituents in grams per cubic centimeter

Humus	1.3–1.5	Anorthite	2.7–2.8
Clay	2.2–2.6	Dolomite	2.8–2.9
Kaolinite	2.2–2.6	Muscovite	2.7–3.0
Orthoclase	2.5–2.6	Biotite	2.8–3.1
Microcline	2.5–2.6	Apatite	3.2–3.3
Quartz	2.5–2.8	Limonite	3.5–4.0
Albite	2.6–2.7	Magnetite	4.9–5.2
Flint	2.6–2.7	Pyrite	4.9–5.2
Calcite	2.6–2.8	Hematite	4.9–5.3

The densities of some of the soil minerals deviate considerably from the average value of 2.65. Those containing substantial amounts of iron are much heavier. The densities of several soil constituents are given in Table 3-2.

Knowledge of the density of soil particles is needed to calculate their sedimentation velocity in water. This is used in erosion research and in mechanical analysis. Density differences are also of value in the separation of the various soil minerals in lithological studies. This separation is done by the use of liquids of high density. Frequently tetrabromoethane (sp. gr. 2.87) or bromoform (sp. gr. 2.89) serve for this purpose. Toluene (sp. gr. 0.86), nitrobenzene (sp. gr. 1.20), or other organic liquids are used to adjust the density to the desired level.

CHEMICAL NATURE

Silica and silicates make up the largest part of the mineral components of the soil. Although the chemical composition of soil particles varies greatly from profile to profile, generally the larger particles are highest in Si; the finer ones contain more K, Ca, and P. The dominant minerals are quartz in sand; quartz and feldspars in fine sand and silt; and mica, vermiculite, montmorillonite, kaolinite, and amorphous colloids in the clay fraction. The larger particles down to the coarse clay are usually broken-down rock fragments, while material below 0.0002 mm is truly colloidal, consisting of clay crystals. Only a small portion of the clay is formed in the soil. The majority is inherited directly from the parent material. Mica and illite, for instance, come from granitic rocks and through weathering give rise to hydrous mica, vermiculite, and montmorillonite.

SPECIFIC SURFACE OF SOIL PARTICLES

The specific surface is probably the outstanding characteristic of soil grains that results from their size. By this is meant the proportion of surface to unit volume of the particles. It is expressed as square centimeters per cubic centimeter or as square meters of surface per gram of soil. Specific surface is important because most chemical and physical reactions occur at the surfaces and therefore the amount of these reactions is approximately proportional to the specific surface. The specific surface of geometric bodies such as cubes or spheres can readily be determined accurately. A cube of 1-cm sides has a specific surface of 6 cm²/cc; a cube of 1-mm sides has a specific surface of 60 cm²/cc. The specific surface of a sphere is

$$\frac{4\pi r^2}{4/3\pi r^3} \qquad \frac{\text{(surface of a sphere)}}{\text{(volume of a sphere)}}$$

This can be canceled to read $3/r$. If we assume $r = 0.62$ cm, the volume of the sphere is 1 cc and its specific surface 4.85 cm²/cc. For spheres of a radius of 0.62 mm in diameter the specific surface is 48.5 cm²/cc. The relationship between the specific surface and the length of the side of a cube can be expressed by the equation

$$S_c = \frac{6}{L}$$

where S_c = specific surface of a cube in cm²/cc
L = side of a cube in cm

The relationship between the specific surface of a sphere and its radius is

$$S_s = \frac{3}{r}$$

where S_s = specific surface of a sphere in cm²/cc
r = radius of the sphere

For the same volume the specific surface of a cube is 1.239 times larger than that of a sphere. In the case of the cube as well as in that of the sphere the specific surface increases tenfold for each tenfold reduction in size. Although soil particles have irregular shapes, they too follow essentially this relationship. This is true of the gravel, sand, silt, and the coarser clay fraction. As clay particles are more nearly plate or lath shaped than spherical or cubical, their specific surfaces are considerably larger than would otherwise be the case. Actually the specific surface of clay is so much larger than that of silt and sand that practically the total surface of a medium- or fine-textured mineral soil is due to the clay. This is particularly true for the expanding clays (see Chap. 4) which have a great deal of internal surfaces between their individual sheets. Assuming a specific surface of 100 m²/g and a bulk

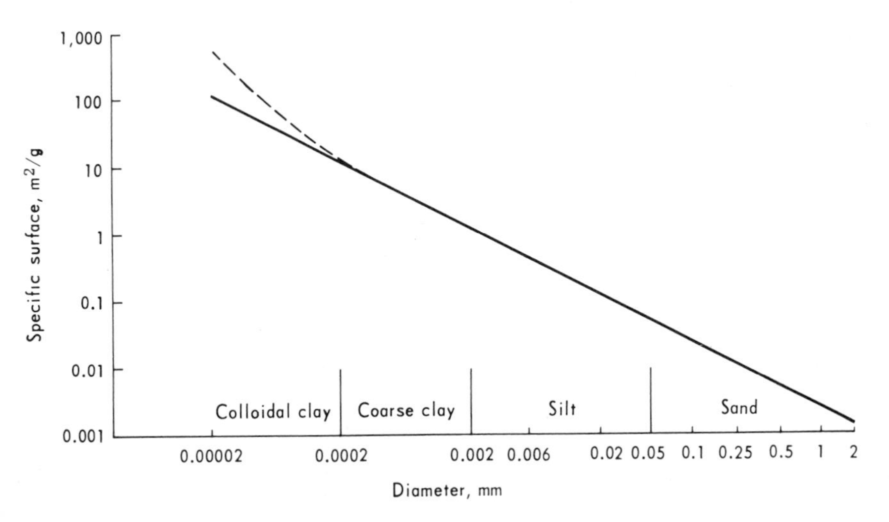

Fig. 3-1 Relationship between particle size of soil fractions and their approximate specific surfaces: The full line represents the relation between the size of spheres of density 2.65 g/cc and their specific surfaces. The broken line takes into account that clay particles are more nearly plate-shaped and that the fine clay has much internal surface.

density of 1.25 g/cc, 1 hectare of a silt loam soil to an ordinary plow depth of 20 cm has a total mineral surface of 25 million hectares (or 100,000 sq miles), a huge area indeed. Also the organic components of the soil, especially the humus, have a very large specific surface as they are colloidal in size.

The specific surface of soil can be estimated from the magnitude of its heat of wetting. It can be determined with a fair degree of accuracy by bringing dry soil in contact with certain organic compounds (e.g., ethylene glycol or glycerol) under specific conditions of vapor pressure that result in a monomolecular layer of these compounds. From the weight increase the total surface of the soil sample can be calculated.

The thickness of hygroscopic water covering the exterior and interior surfaces of soil particles is an approximate function of the tension. It is therefore possible to estimate the specific surface of a soil from the water it holds at a definite tension in the hygroscopic range.

Expressing the mechanical composition of soils as the specific surface may be more meaningful than as texture, since the physical and chemical reactions of the soil largely take place at the surfaces. It has the advantage of using a single figure (square meters per gram) to describe the mechanical composition. Table 3-3 gives typical ranges of specific surface for clays and for soils of different textures.

Table 3-3 Ranges of specific surface of clays and soils in square meters per gram

Clay, montmorillonite	500–800
Clay, illite	60–120
Clay, kaolinite (soil)	20– 40
Clay, kaolinite (ceramic)	10– 20
Clay, soil-clay fraction in Middle West	300–400
Clay, from glacial calcareous till	150–200
Clay soil	150–250
Silty clay loam	120–200
Silt loam	50–150
Loam	50–100
Sandy loam	10– 40
Silt soil	5– 20

SOIL TEXTURES

The relative proportion of the various ultimate grain-size fractions in the soil is called *soil texture*. Texture designations are made up by using the names of the predominant size fractions and the word "loam" whenever all three major size fractions occur in sizable proportions. Thus the term "silty clay" describes a soil in which the clay characteristics are outstanding and which also contains much silt. A "silty clay loam" is similar to the silty clay except that it contains more sand, and therefore is somewhat more mellow.

Frequently also the fineness of the sand is mentioned in the texture name, such as "very fine sandy loam." The U.S. Department of Agriculture has developed a system of texture designation that is based on that Department's soil-particle-fraction classification. This system is generally accepted in the United States and several other countries. The 12 main textures and their compositions are shown on the U.S. Department of Agriculture soil-texture triangle (Fig. 3-2). A glance at the texture triangle indicates the importance of specific surface. It takes more than 80 percent of silt to call a soil a "silt," and more than 85 percent of sand to call a soil a "sand," but only 40 percent clay is required to call a soil a "clay." When dealing with the mechanical composition of soil one has to realize that the terms "clay," "sand," and "silt" are used both for soil separates and for texture designations. The percentage of the individual fractions is calculated on the basis of organic-free, oven-dry soil particles less than 2 mm in diameter.

Although the organic matter content in mineral soils (soils with more than about 85 percent mineral matter) is not expressed in the U.S. Department of Agriculture texture designation, it is of great impor-

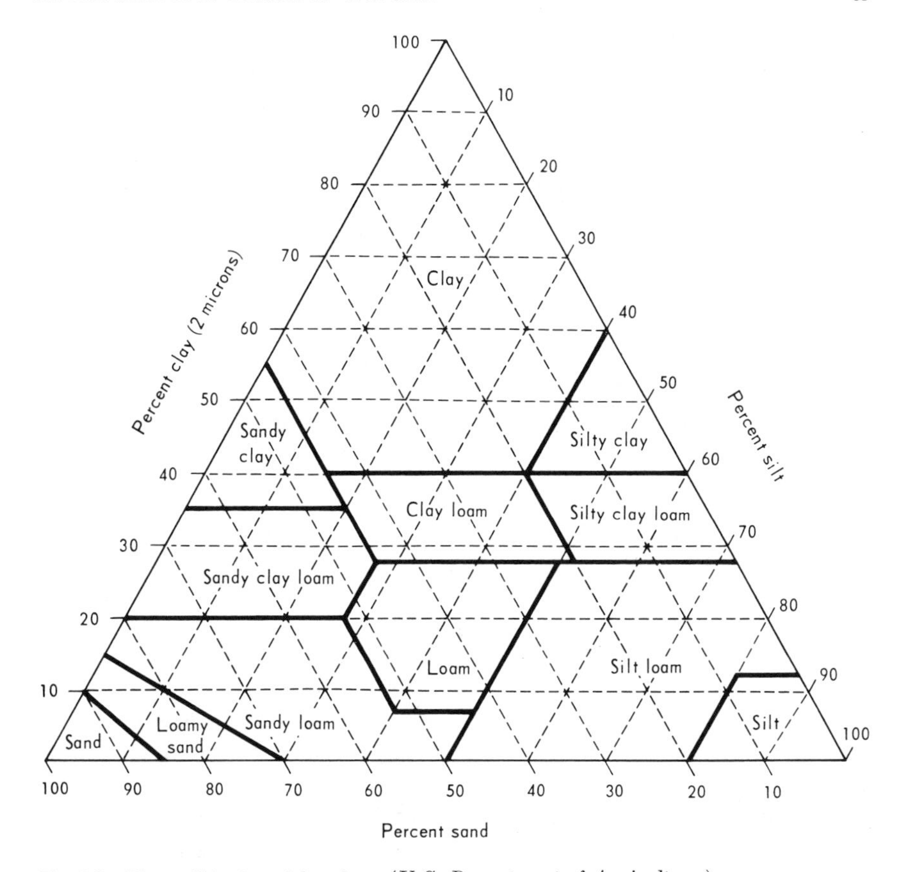

Fig. 3-2 The soil-texture triangle. *(U.S. Department of Agriculture.)*

tance in determining the value of the soil. The geographic soil series name serves to imply the organic matter content to the person familiar with the local soil classification. It seems that it would be preferable to develop a more direct system of nomenclature that would denote the organic matter content. Soils containing over 15 percent organic matter are designated as mucky or muck, for well-decomposed material, and peaty or peat, for soils with only partially decomposed plant residues.

A knowledge of the texture of a soil is of obvious importance. It is a guide to the value of the land. Land use capability and methods of soil management are largely determined by the texture. Generally the best agricultural soils are those that contain 10 to 20 percent clay, 5 to 10 percent organic matter, and the rest divided about equally between sand and silt. In the study of the morphology and genesis of soils and for their classification and mapping, an accurate knowledge of the texture

Table 3-4 Soil-texture classification

Coarse-textured soils	1. Sands and loamy sands	Sand Loamy sand
Medium-textured soils	2. Sandy loams	Sandy loam Fine sandy loam
	3. Loamy soils	Very fine sandy loam Loam Silt loam Silt
	4. Moderately heavy soils	Sandy clay loam Clay loam Silty clay loam
Fine-textured soils	5. Clays and silty clays	Sandy clay Silty clay Clay

of soils is imperative. It must be kept in mind that the textures of the entire profile are of importance, not only that of the surface layer.

It has been proposed to group the various soil textures in the manner shown in Table 3-4.

Formerly coarse-textured soils have been called "light" soils and fine-textured soils "heavy" soils. This was based on the relative ease of draft required to pull tillage equipment, especially the plow, through the ground. These terms have been abandoned because they tend to be confusing.

The following definitions of the basic soil-textural classes are essentially those given by the U.S. Department of Agriculture (1951 and 1960).

SAND

Sand is loose and single-grained. The individual grains can readily be seen or felt. Squeezed in the hand when dry it will fall apart when the pressure is released. Squeezed when moist, it will form a cast, but will crumble when touched.

SANDY LOAM

A sandy loam is a soil containing much sand but which has enough silt and clay to make it somewhat coherent. The individual sand grains can readily be seen and felt. Squeezed when dry, it will form a cast which will readily fall apart, but if squeezed when moist a cast can be formed that will bear careful handling without breaking.

LOAM

A loam is a soil having a relatively even mixture of different grades of sand and of silt and clay. It is mellow with a somewhat gritty feel, yet fairly smooth and slightly plastic. Squeezed when dry, it will form a cast that will bear careful handling, while the cast formed by squeezing the moist soil can be handled quite freely without breaking.

SILT LOAM

A silt loam is a soil having a moderate amount of the fine grades of sand and only a small amount of clay, over half of the particles being of the size called "silt." When dry it may appear cloddy but the lumps can be readily broken, and when pulverized it feels soft and floury. When wet the soil readily runs together and puddles. Either dry or moist it will form casts that can be freely handled without breaking, but when moistened and squeezed between the thumb and finger it will not "ribbon" but will give a broken appearance.

CLAY LOAM

A clay loam is a fine-textured soil which usually breaks into clods or lumps that are hard when dry. When the moist soil is pinched between the thumb and finger it will form a thin "ribbon" which will break readily, barely sustaining its own weight. The moist soil is plastic and will form a cast that will bear much handling. When kneaded in the hand it does not crumble readily but tends to work into a heavy compact mass.

CLAY

A clay is a fine-textured soil that usually forms very hard lumps or clods when dry and is quite plastic and usually sticky when wet. When the moist soil is pinched out between the thumb and fingers it will form a long, flexible "ribbon." Some clays, especially the latosolic ones in the tropics, are friable and lack plasticity in all conditions of moisture.

The three main soil components can be recognized in the field by these simple tests:

Sand feels gritty,
Silt, when pressed with a finger, shows fingerprints,
Clay, when streaked out moist between two thumbnails, shows a shiny surface.

The significance of soil textures is summarized in Table 3-5.

Table 3-5 Significance of soil textures

	Sand	*Loam*	*Silt loam*	*Clay soil*
Feel	Gritty	Gritty	Silky	Cloddy or plastic
Identification	Loose	Cohesive	Shows fingerprint	Gives shiny streak
Internal drainage	Excessive	Good	Fair	Fair to poor
Plant-available water	Low	Medium	High	High
Drawbar pull	Light	Light	Medium	Heavy
Tillability	Easy	Easy	Medium	Difficult
Runoff potential	Low	Low-medium	High	Medium-high
Water detachability	High	Medium	Medium	Low
Water transportability	Low	Medium	High	High
Wind erodibility	High	Medium	Low	Low

The deeper penetration of roots in the coarse-textured soils compensates to some extent for their lower holding capacity for available water. Under the standard conditions of the U.S. Erosion Experiment Stations Olson and Wischmeier (1963) found the erodibility of soils to increase in the following sequence of textures:

Flaggy and gravelly soils,
Loamy sand,
Sandy loam,
Clays and clay loams,
Silt loam.

MECHANICAL ANALYSES

Definition The determination of the relative distribution of the size groups of ultimate soil particles is called *mechanical analysis*. In all types of quantitative mechanical analyses two steps are necessary.

1. Separation of all particles from each other—complete dispersion into ultimate particles.
2. Measuring the amounts of each size group in the sample.

DISPERSION OF THE SOIL SAMPLE

To separate the ultimate soil particles from each other we have to know what holds them together. The following table shows the agents holding soil together and the means to overcome this action in each case.

Aggregating agents	*Methods of dispersion*
Surface tension	Elimination of air by stirring sample in water or boiling it, or elimination of water (as it is done for sieving sand)
Lime and colloidal oxides of iron and aluminum	Dissolving the aggregating agent in dilute HCl
Organic matter	Oxidize with H_2O_2
High concentration of electrolytes	Precipitate and decant or filter with suction
Too low electrokinetic potential	Remove or complex polyvalent cations and replace with monovalent cations, especially Li and Na. (Sodium metaphosphate or sodium carbonate are frequently used.)

No matter what method of mechanical analysis is used, it can be accurate only if the dispersion is complete.

In order to check the effectiveness of the dispersion methods, the composition of a soil mixture of known proportions of sand, silt, and clay is determined after employing the various methods of dispersion.

MEASURING PARTICLE-SIZE FRACTIONS

Once the soil sample is dispersed into its ultimate particles, several methods can be used to measure the proportions of each particle-size fraction. Generally, the coarser fractions are separated by sieving and the finer ones by settling in a viscous medium.

Sieving The sieves used in mechanical analysis correspond to the desired particle-size separation. For the sizes of 2 mm, 1 mm, and 0.5 mm, generally sieves with circular holes are employed. For the smaller sizes wire mesh screens are used. Since the thickness of the wires for screens of the same number of mesh per inch is not the same in all screens, there is no definite relationship between the number of mesh per inch and the size of the screen openings. Generally 60 or 70 mesh to the inch corresponds to 0.25-mm openings, 140 mesh to the inch corresponds to 0.1-mm openings, and 270 or 300 mesh to the inch correspond to 0.05-mm openings. This is the smallest opening normally used in soil screening. This means that screening is used to separate the sand into its size fractions. Screening may be done with dry or wet soil. If it is to be dry, most of the clay should be first removed by decantation, as otherwise an ultimate separation would not be effected. When it is to be done wet, special precautions have to be taken to overcome the surface tension below the finest screens.

After wet sieving, sand should be dried and sieved again, because usually a part of the silt fraction will be caught on the 0.05-mm screen when the sieving is done wet.

Sedimentation It has long been known that the velocity of fall of an object will be influenced by such conditions as

The viscosity of the medium.
The difference in density between the medium and the falling object. In fact in some cases the suspended object actually rises—like oil in water, or gas in a liquid.
The size and the shape of the object.

In 1851, Stokes formulated the law governing the rate of settling of spherical particles in a viscous medium. It states that the resistance offered by a liquid to the fall of a rigid spherical particle varies with the circumference of the sphere, not with its surface. On the other hand, the force of fall by the particle is proportional to its weight, and consequently to its volume. The components that make up the equation of Stokes' law are shown as contributing to the cause of settling and to the resistance to settling.

$(4/3)\pi r^3$ ·	$(d_1 - d_2)$ ·	g	=	$2\pi r$ ·	z ·	$3V$
Volume of particle	Density difference	Acceleration		Circumference of particle	Viscosity	Velocity of sedimentation

Cause of settling = resistance to settling

$$V = \frac{(4/3)\pi r^3(d_1 - d_2)g}{2\pi r z 3} = \frac{2r^2(d_1 - d_2)g}{9z}$$

or

$$V = Kr^2$$

where V = velocity of fall, cm/sec
d_1 = density of particle, g/cc
d_2 = density of liquid, g/cc (Stokes' law does not hold for gases as fall is too rapid and air itself is moved.)
g = acceleration, cm/sec^2. This may be due to gravity or due to centrifugal force.
r = radius of particle, cm
z = absolute viscosity, poises
K has the dimensions of $L^{-1}T^{-1}$

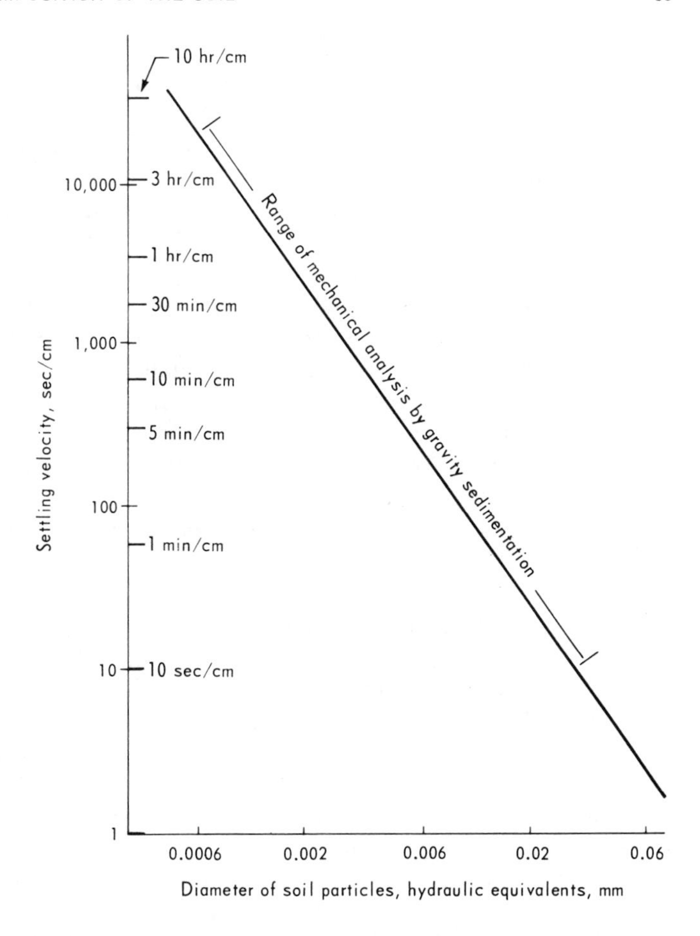

Fig. 3-3 Settling velocities of soil particles in water: Temperature, 25°C; gravitation, 980.15 g/cm-sec^2; density, 2.65 g-cc.

The relationship between the size of a soil particle and its velocity of sedimentation in water is illustrated in Fig. 3-3 assuming specific conditions of temperature, particle density, and gravitational force. It shows that decreasing the diameter of a sphere to one-tenth its size decreases its settling velocity 100-fold.

Conditions under which Stokes' law is valid (assumptions in the equation) are

1. Particles must be spherical and rigid. Particles of other shapes fall faster or slower. In practical mechanical analysis hydraulic equivalents of spheres are used without regard to their actual sizes and

shapes. The noncolloidal soil particles are not far from being spherical. Colloidal particles are plate-shaped and fall slower than spherical particles of the same mass.

2. Particles must be large in comparison with the molecules of the liquid, so that in comparison with the particle the medium can be considered homogeneous, i.e., particles must be big enough to have no brownian movement if acceleration due to gravity is employed. The larger limit of particles exhibiting brownian movement is approximately 0.0002 mm. In the gravitational field particles from very fine sand to coarse clay can be determined with the sedimentation method. With a centrifuge soil particles down to 0.00006 mm can be separated.
3. Fall must be unhindered. Particles falling very near the wall of the vessel (0.1 mm in distance) are slowed down in their descent. Particles must not hinder each other. Many fast-falling particles may drag finer particles down along with them. The concentration may be up to 5 percent solids in the suspension, preferably less than 3 percent.
4. Particles must be of uniform density. After elimination of organic matter the soil particles vary in density from 2.5 to 3.4, but the majority of the mineral particles in most soils vary only from 2.6 to 2.75, and an average of 2.7 can be used with reasonable accuracy as a basis for calculations. Note that a determination of the density of the soil containing organic matter is valueless for this purpose.
5. The suspension must be still. Any movement of the suspension will alter the velocity of fall. Such movement can be brought about by the original mixing of the soil with the liquid, by heat convection, or by sedimentation of large particles. Particles larger than 0.08 mm settle so fast that they set up turbulence and consequently their rate of fall cannot be calculated from Stokes' law.

SPECIFIC METHODS OF MECHANICAL ANALYSIS

Numerous methods for laboratory and field use have been developed throughout the years. These include the elutriator method, the test-tube-shaking method together with some refinements of the technique, the Wiegner sedimentation cylinder method that is based on the different densities of soil suspension, the photoelectric method, the pipette method, and the hydrometer method. Only the two latter methods have found general acceptance. The pipette method is preferred for accurate work, while the hydrometer method is considerably faster and still accurate enough for most purposes. The foregoing conditions show that definite precautions have to be taken to use sedimentation for mechanical analysis.

In the pipette method a sample of the soil suspension is taken at a given depth below the surface at a predetermined time. In this way the sample contains all the fractions still in suspension at that depth. By sampling at different intervals after the stirred suspension has come to rest the mechanical composition of the soil can be calculated. For water at 25°C and soil particles of 2.65 g/cc density in the gravitational field (980.15 g/cm-sec²), Stokes' law becomes $V = 40285r^2$. This means that a particle of 0.05 mm in diameter falls at a velocity of 0.25 cm/sec. That is the same as 5 cm in 20 sec. It is clearly impossible to take an accurate sample of the suspension with a pipette 20 sec after the suspension has come to rest. The beginning of the sedimentation period after the suspension has been stirred cannot be told for certain for less than 2 or 3 sec. The sampling with the pipette requires 10 sec after it is placed at the correct elevation. The sampling of the silt plus clay fraction is therefore only an approximation. For particles of 0.01 mm in diameter, the rate of sedimentation is 0.01 cm/sec or 5 cm in 8 min and 20 sec. This period makes it possible to sample fine silt plus clay in a satisfactory manner by the pipette technique.

The use of the hydrometer for mechanical analyses of soil, originally suggested by Bouyoucos (1927), is based on the continuous reduction of the density of the soil suspension with time at the rate the particles drop below the level of the hydrometer. At any one moment the density of the suspension is lowest near the top and increases toward the bottom.

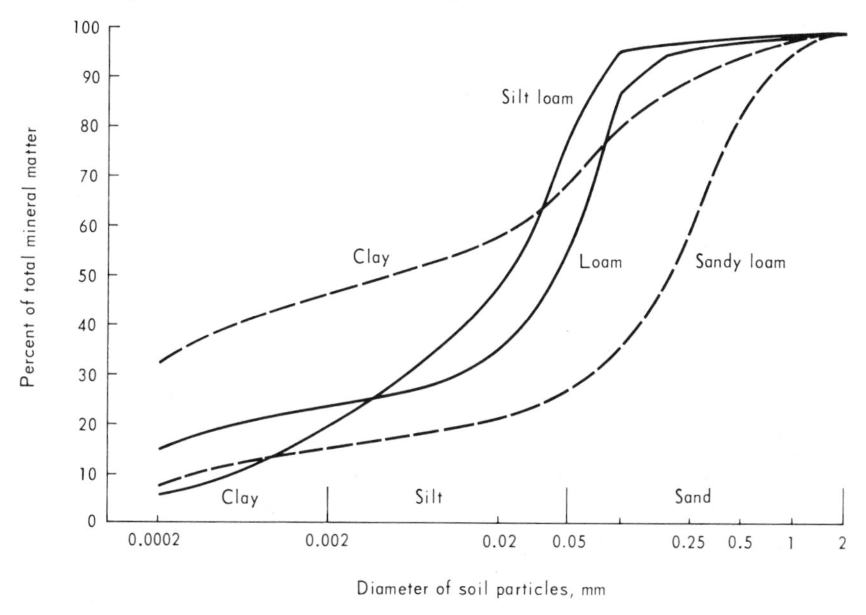

Fig. 3-4 Mechanical composition of typical soils: summation percentage graph.

Table 3-6 Soil mechanics spectrum

Soil mechanics spectrum

Soil fraction	Diameter			Methods of investigation	Methods of mechanical analysis
	mm	microns	Å		
Gravel					
	2	2000			
Coarse sand					
					Screening
	0.2	200			
Fine sand				Optical	
	0.05	50			
				Microscope	Problem zone
	0.02	20			
Silt					Sedimentation by gravity
	0.002	2	20,000	X-ray diffraction Electron microscope	
Coarse clay		Red*	7000		
	0.0005		5000	Limit of optical microscope	
		Violet*	4000		
	0.0002	0.2	2000	Infrared spectroscopy	
					Sedimentation by centrifuge
	0.00006	0.06	600		
Colloidal clay					
	0.00002	0.02	200	Limit of x-ray diffraction and of electron microscope	
	0.000005	0.005	50		
	0.000002	0.002	20	Limit of infrared spectroscopy	
True solution					

*Wavelength of visible light

This has to be taken into account in the calibration of a hydrometer for the determination of the amount of soil in suspension.

REPORTING THE MECHANICAL COMPOSITION

Once the amounts of sand, silt, and clay in a soil have been determined, the results can be reported in various ways. They may merely be stated as a percentage in tabular form. They may be shown as a point on the texture diagram. This makes it possible to state the texture of the analyzed soil. They may be plotted on 5- or 6-cycle semilog graph paper with the percentage of samples as the ordinate on the linear scale and the diameter of the particles as the abscissa on the logarithmic scale. This is called a *summation percentage graph.* Examples of plotting mechanical-composition data for several soils are shown in Fig. 3-4.

Table 3-6 is presented to illustrate the relationships between the various soil fractions and the methods used for their investigation and mechanical analyses.

REFERENCES

Bouyoucos, G. J.: The Hydrometer as a New Method for the Mechanical Analysis of Soils, *Soil Sci.*, vol. 23, pp. 343–353, 1927.

Olson, T. C., and W. H. Wischmeier: Soil-erodibility Evaluations for Soils on the Runoff and Erosion Stations, *Soil Sci. Soc. Am. Proc.*, vol. 27, pp. 590–592, 1963.

Soil Survey Staff: "Soil Survey Manual," *U.S. Dept. Agr. Handbook* no. 18, 1951.

———: "Soil Classification, a Comprehensive System," 7th Approximation, Soil Conservation Service, U.S. Department of Agriculture, 1960.

Wadell, H.: Volume, Shape and Roundness of Rock Particles, *J. Geol.*, vol. 40, pp. 443–451, 1932.

4 SOIL CLAYS

THE NATURE OF CLAY

The term *clay* is used to designate different things by different people; but in soil science, clay is that fraction of the mineral components of soil which are smaller than 0.002 mm in any dimension. This includes layered aluminosilicate crystals, oxides of iron, aluminum, and silicon, as well as finely comminuted particles of any minerals that occur in the soil and also any amorphous mineral particles in this size group. Soil scientists also refer to a soil that contains more than 40 percent clay as "clay." A similar definition is also used in geology. This meaning will not be used in this chapter.

Much of the soil clay is colloidal. This means that it is made up of particles that are so small that distinctive reactions of the surface are appreciable. There is no sharp line between colloidal and noncolloidal clay, but frequently the boundary is arbitrarily set at 0.0002 mm. Colloids in suspension exhibit brownian movement. That means they move back and forth as the result of the impacts of the molecules of the surrounding medium. In the case of larger particles the impacts from all sides compensate each other and the particle remains stationary.

Originally all clay was assumed to be amorphous. Most of the research before 1925 was based on that assumption. However, certain observations have since shown that at least some of the clay must have planar surfaces:

Scintillating or twinkling of clay suspensions is visible with the naked eye and in the ultramicroscope.
Most clay, when smoothed out with a flat tool, has a reflective surface.
Thin layers of clay curl up when dried.
Clays show entirely different exchange properties than mixtures of colloidal oxides of silicon, aluminum, and iron made up to resemble the various clays in the chemical composition.
Studies with the electron microscope and with x-ray diffraction have since given clear evidence that many soil clays are made up of layered crystals.

This chapter deals with the subject of clays to the extent it is of specific interest in the study of soil physics. For more detailed information the following books are recommended: Grim (1968), Marshall (1964), Rich and Kunze (eds.) (1964), and Black (ed.) (1965) (see References).

PROPERTIES OF THE LAYERED ALUMINOSILICATE CLAYS

The layered aluminosilicate clay crystals probably constitute the most important fraction of soil clay. They have been studied more than the other fractions and more is known about their effects on soil characteristics. It is also true that they possess the properties that are generally ascribed to clay to a much higher degree than the other components of "soil clay."

CRYSTAL STRUCTURE

Each type of clay has a specific crystal structure. Although the chemical composition of clays is variable, some generalizations are nevertheless possible. The layered aluminosilicate clay crystals are made up of sheets of tetrahedra of silicon oxide and of sheets of octahedra of oxide and hydroxide of aluminum. This means that the sheets are made up of four-sided and eight-sided molecular building blocks, respectively (Fig. 4-1).

A tetrahedral shape is formed when a small atom like silicon is surrounded by four oxygens. An octahedral shape is formed when six oxygens or OH groups surround a larger atom like aluminum.

ISOMORPHIC REPLACEMENT

Sometimes aluminum replaces silicon in the tetrahedra while iron, magnesium, manganese, or a few other cations of similar size may replace aluminum in the octahedra. Since the replacing ions in these *isomorphic substitutions* (substitutions that maintain the structure) are larger than the original ions, strains are set up in the lattice and prevent the crystals

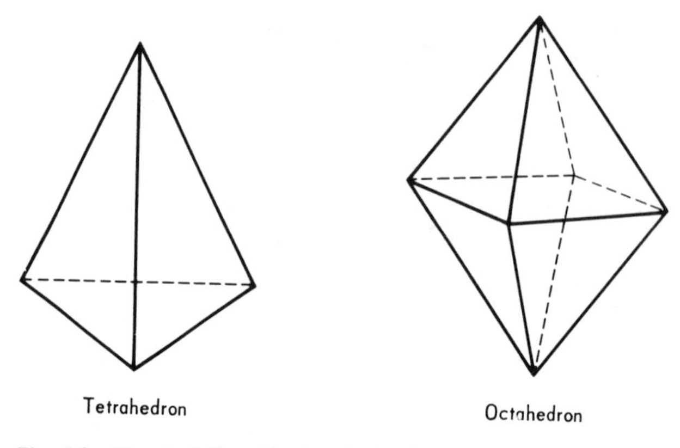

Fig. 4-1 The building blocks of aluminosilicate clays.

from growing too large. This is the main reason that most clay crystals do not reach the size where they can be seen with an optical microscope. In general one ion may substitute for another if their ionic radii differ by less than 15 percent.

It must be stated here that this "replacement" or "substitution" takes place at the time the clay is originally formed. After clay formation, removal of one ion from the clay crystal and the replacement by another is very slow, except for the *exchangeable cations* at the surfaces of the clay.

EXCHANGEABLE CATIONS

The valence of the replacing positively charged ions in many cases is lower than that of the ions normally present. This means that some of the negative valences of the oxygen are not satisfied internally and bring about an overall net negative charge for these clays. As a consequence such clays function similarly to weak acids. This total negative charge has to be neutralized by cations. In the case of the diphormic clays and the expanding triphormic clays these cations can be exchanged readily. In the nonexpanding or slightly expanding triphormic clays some of the interlayer sites are inaccessible from outside the clay crystal. The cations located in such positions can be exchanged only after partial disintegration of the clay. Consequently the cation exchange capacity of such clays is smaller than their total negative charge. Both these properties are usually expressed in milliequivalents of cations per 100 g of dry clay.

Because the seat of the negative valences is at different distances from the surface, the energy with which they hold the cations is different. Therefore, the cation exchange capacity (CEC) as measured depends on the bonding energy of the exchangeable cation and the concentration of the cation. Thus, the cation exchange capacity is influenced by the method of determination used.

In addition to the exchange capacity brought about by isomorphic substitution, electric charges are developed at the edges of the clay crystals because of broken bonds. These charges may be positive or negative. All these charges contribute to the ability of clays to hold ions. This ability is of utmost importance to the nutrition of plants and microbes, as these ions can be readily given off and exchanged for others, e.g., hydrogen. Generally the optimum condition for plant nutrition is reached when about four-fifths of this exchange capacity is saturated with bases and the rest with hydrogen.

DISPERSION AND FLOCCULATION

Since the force with which the cations surrounding the clay crystal are held decreases with the square of the distance between the cation and the negative sites, some of the ions are held very close to the surface and they neutralize the negative charges effectively, while the more hydrated cations, e.g., sodium, remain at a somewhat greater distance and neutralize only part of the negative charge. Under this condition the effective net charge remains so large that the clay particles repel each other and, if they are in aqueous suspension, will stay dispersed indefinitely.

Divalent ions, such as calcium and magnesium, are also hydrated, but their higher charge causes them to be held more closely to the negatively charged clay. When they saturate the clay, its zeta potential—its negative electric charge—is small and the individual gravitational attraction (van der Waals' forces) between the particles is sufficiently large to overcome the repulsion due to the electric charges. The result is that clay particles in water will group themselves in large numbers; they will become *flocculated.*

SWELLING AND SHRINKING

One of the outstanding properties of some clays of the montmorillonite group is their ability to swell on wetting and to shrink on drying. These clays take up water between the sheets of ions that form the crystal and increase in volume correspondingly. The degree of swelling depends upon the specific type and particle size of the clay and is a function of the size and valence of the exchangeable cation. Sodium clays, for instance, swell and shrink much more than calcium clays. Most other clay types swell only little or not at all.

SPECIFIC SURFACE

Due to its small particle size, clay has a large specific surface. This ranges from 15 to 800 m^2/g from the coarsest nonexpanding clays to the finest expanding clays. In the case of the swelling clays, the surfaces between the individual sheets contribute the largest portion of the specific surface. Clay has a high heat of wetting. This is somewhat proportional to the specific surface. In fact it is possible to estimate the specific surface and indirectly the percentage of clay in a soil devoid of organic matter by determination of the heat of wetting, as neither silt nor sand have any appreciable heat of wetting.

Due to their strong absorptive power for water, clays take up much water even from an atmosphere only half saturated with water vapor. This makes it possible to estimate the proportion of colloid in a soil from the moisture held at 50 percent relative humidity, since the hygroscopic water capacity of silt and sand is exceedingly small. If such a soil contains only a small amount of organic matter we can conclude that it is high in colloidal clay.

The small size of clay crystals together with their plate shape bring about a very high cohesiveness that contributes much to soil-structure formation. This is particularly true for the swelling clays.

SIZE RELATIONSHIPS OF THE ATOMS IN CLAY CRYSTALS

For the understanding of the construction of the various types of clay it is important to know the size relationships of the atoms making up the building blocks of the clays. The ionic radii of the more important ions occurring in clays are given in Table 4-1.

All cations occurring in clays are smaller than the oxygen ion. The most important cations in clays are silicon and aluminum. Silicon is small enough (radius 0.41 Å) to fit into the pore space of an oxygen tetrahedron (effective radius 0.30 Å) without pushing the oxygens far apart. It must be realized that ions are not like small marbles with discrete diameters but that circumferences of their electron paths are flexible. When an aluminum ion (radius 0.45 Å) is located in the tetrahedral space of oxygens or hydroxyl groups, these are considerably sep-

Table 4-1 Ionic radii of ions occurring in crystalline clays

Ion	*Radius, Å*	*Ion*	*Radius, Å*
O^{--}	1.45	Li^{+}	0.68
K^{+}	1.33	Mg^{++}	0.65
Na^{+}	0.98	Ti^{4+}	0.64
Ca^{++}	0.94	Fe^{3+}	0.53
Mn^{++}	0.80	Al^{3+}	0.45
Fe^{++}	0.75	Si^{4+}	0.41
Zn^{++}	0.70	B^{3+}	0.20

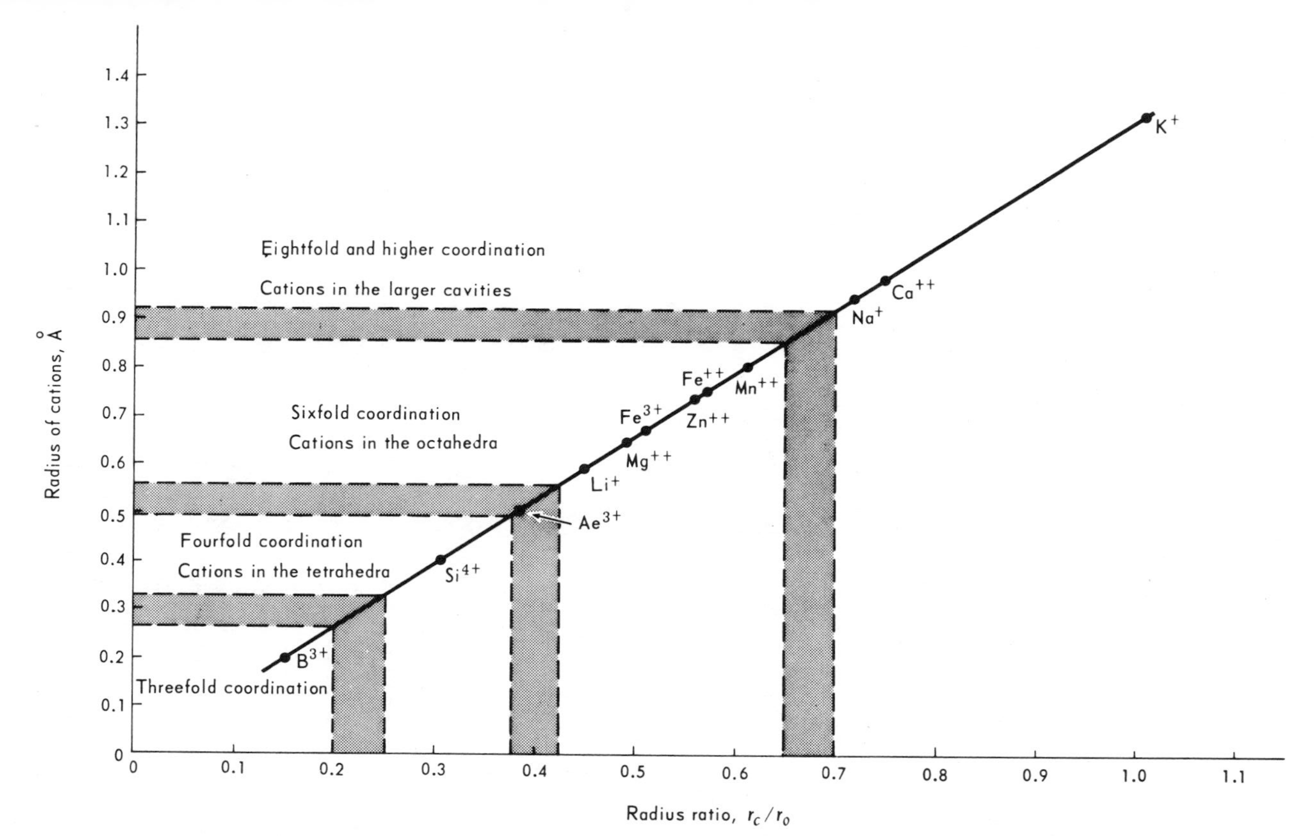

Fig. 4-2 Coordination of cations with oxygen ions in clays: The broad boundary bands indicate that there is no sharp separation between the different coordination levels.

arated and do not hold together as much as when a silicon is in the center of the tetrahedron.

Similarly aluminum fits well into the space of an oxygen octahedron. This space has an effective radius of 0.55 Å. The ions that replace aluminum, for instance, ferrous iron, magnesium, and manganese, are considerably larger than the space of a tightly fitting oxygen octahedron. The result is that the oxygen atoms are more widely separated. This replacement of cations both in the tetrahedra and in the octahedra is therefore the reason that the lattice of such clay crystals has little stability and the individual clay particles are always small.

A cation located in the center of a tetrahedron has four nearest anion neighbors. This is called *fourfold coordination*. A cation located in the center of an octahedron has six nearest anion neighbors. This is sixfold coordination. In the same way eightfold and higher coordination exists in the center of more complex ionic structures. Whether a cation occurs in fourfold or sixfold or any other coordination depends on the relative size of the cation with respect to the surrounding anions. This relationship is normally expressed as the *radius ratio*, r_c/r_o, the ratio of the radius of the cation to the radius of the anion. Figure 4-2 shows the radius ratios, the ionic radii as well as the oxygen coordination numbers of several cations occurring in clays. It will be noted that ions with a radius ratio of more than 0.65, e.g., potassium and sodium, are too large to fit into either tetrahedra or octahedra. They find sufficient space only in openings of eightfold or higher coordination.

THE MAIN GROUPS OF CLAY CRYSTALS

Although all clays are similar in some of their important properties, there are many types of clays and they can be classified in various ways. As previously stated, crystalline clays are made up of sheets of octahedra and tetrahedra of oxygen and hydroxyls.

The tetrahedra in themselves hold each other very tightly and also the octahedra hold each other very tightly, but the attraction between pairs of tetrahedral sheets is somewhat weaker. This is the reason that clay crystals generally cleave along flat planes. Some clays are made up of alternating sheets of tetrahedra and octahedra. The ratio of tetrahedral sheets to octahedral sheets is 1:1. They are called 1:1 clays or *diphormic clays* (Fig. 4-3).

In another group of clays the unit layer is made up of two sheets of tetrahedra enclosing a sheet of octahedra. The ratio of tetrahedra to octahedra sheets is thus 2:1. In this way two tetrahedral sheets are lying next to each other. Since these two identical sheets do not attract each other by chemical bonds, but only by relatively weak van der Waals'

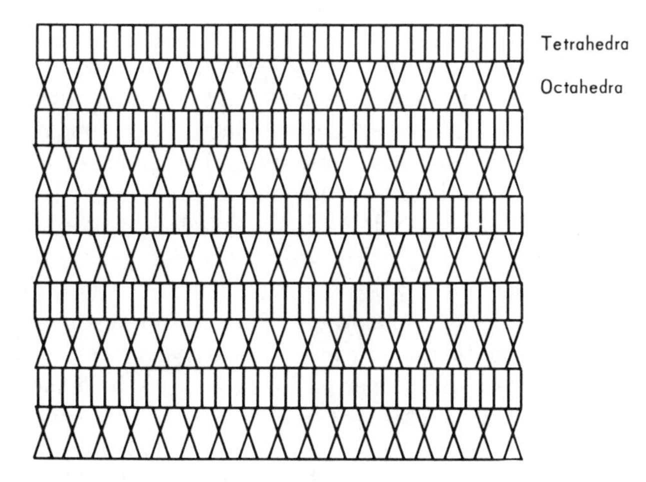

Fig. 4-3 Diphormic clay: schematic cross section.

forces, they can be separated by water molecules. Such clays therefore have the ability to swell on wetting and to shrink on drying. The montmorillonite is the best-known member of this type of clay. These are 2:1 clays or *triphormic clays* (Fig. 4-4).

In the case of some of the 2:1 clays, potassium or magnesium atoms are located between the neighboring tetrahedral sheets and hold these

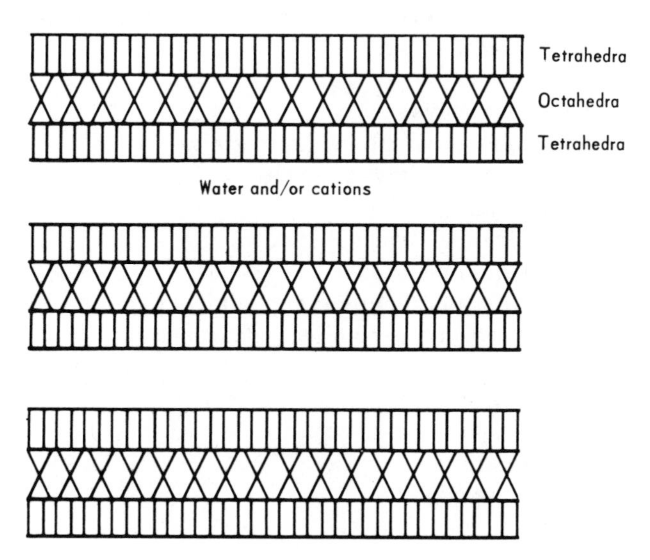

Fig. 4-4 Triphormic clay: schematic cross section.

sheets tightly together with electrostatic bonds. Consequently these clays do not shrink or swell. The typical clay of this group is muscovite.

Clay crystals can be considered to be made up of individual building blocks that are placed side by side in all three dimensions. Each of these building blocks is called a *unit cell.* The chemical and structural description of a clay is based on the unit cell.

The oxygens and the hydroxyls that make up the corners of the octahedra are shared with neighboring octahedra and to a certain extent also with neighboring tetrahedra. The geometry of the octahedra is such that there are six voids in each unit cell or three voids in each half-unit cell. To satisfy the valences of the oxygens and hydroxyls, six positive valences are required per half-unit cell. Where these valences are provided by divalent cations (Fe^{++}, Mg^{++}, Mn^{++}), each of these three voids in the octahedra is filled with a cation. This arrangement is called *trioctahedral clay.* Where trivalent cations (Fe^{3+}, Al^{3+}) are filling the voids, only two cations per half-unit cell are needed to satisfy all the valences. Therefore, only two-thirds of the vacant sites are filled with cations. Such a mineral is called *dioctahedral.* Dioctahedral sheets are called *gibbsite* sheets. Trioctahedral sheets are called *brucite* sheets. $Al_2(OH)_6$ and $Mg_3(OH)_6$ are the typical compositions of the gibbsite and the brucite sheets, respectively.

CLASSIFICATION OF CLAYS

Using these various characteristics, crystalline clays can be classified in the following manner:

1. Two-layer clays
   a. Equidimensional, kaolinite group
   b. Elongate, halloysite group
2. Three-layer clays
   a. Expanding lattice
      i. Equidimensional
         Montmorillonite group—full expansion
         Vermiculite—limited expansion
      ii. Elongate
         Nontronite, saponite, hectorite
   b. Nonexpanding lattice
      Illite group
3. Four-layer clays
   Chlorite group
4. Chain-structure clays
   Attapulgite group

Each of these groups and each clay type in each of these groups differs from the others in some respect. These properties are described in detail in the specialized clay crystallography literature. Here only three of the most important clay groups occurring in the soil are described (Table 4-2).

Table 4-2 Chief characteristics of the main groups of crystalline clays

	Two-layer clays, 1:1, *"diphormic"*	*Three-layer clays,* 2:1, *"triphormic"*	
		Expanding lattice	*Nonexpanding lattice*
Group members	*Equidimensional* kaolinite, dickite, nacrite *Elongate* halloysite	*Equidimensional* montmorillonite, sauconite, beidellite, vermiculite *Elongate* nontronite, saponite, hectorite	*Equidimensional* illite, biotite, muscovite
Structure	*Rigid lattice* one tetrahedral layer for each octahedral layer	Lattice expanding and contracting with water content. Sheets made up of one octahedral layer with a tetrahedral layer on both sides	*Rigid lattice* Sheets made up of one octahedral layer with a tetrahedral layer on both sides. K bonds sheets together
Swelling and shrinking with moisture changes	Very little	Much	*Illite* little *Micas* none
SiO_2/R_2O_3	1.9–2.1	4.0–6.0	2.1–3.2
Ionic (isomorphic) substitution in lattice	None	Al for Si; Mg, Fe, Mn, Ti for Al; Li for Mg	Al for Si
Adsorptive capacity for inorganic and organic ions, water, gases	Small	Large	Intermediate
Total negative charge	5–10 meq/100 g	80–120 meq/100 g	About 250 meq/100 g
Cation exchange capacity	5–10 meq/100 g	80–120 meq/100 g	*Illite* 20–40 meq/100 g *Micas* 2–5 meq/100 g
Stability of H (Al) clay in aqueous suspension	Low	Great	Intermediate
Heat of wetting	1–2 cal/g	10–20 cal/g	Around 4 cal/g
Specific surface	25–50 m^2/g	Around 750 m^2/g	75–125 m^2/g
C axis basal spacing, repeat thickness with ethylene glycol	7.2 Å	*Montmorillonite* 17 Å *Vermiculite* 14 Å	10 Å
Temperature of first endothermic depression	600°C	100–200°C	*Illite* 50–100°C *Biotite* 100–400°C *Muscovite* 900°C

CLAYS THAT ARE NOT LAYERED ALUMINOSILICATES

In addition to the layered aluminosilicates the clay fraction of soils includes noncrystalline aluminosilicates, such as allophane, as well as oxides and hydroxides of aluminum, iron, and silicon and any rock or mineral fragments that are smaller than 0.002 mm.

Allophane is a general term for amorphous aluminosilicate gels of a wide range of composition. In common with layered aluminosilicate clays, allophanes have silicon in tetrahedral coordination and aluminum in octahedral coordination. But the arrangement of the tetrahedra and octahedra is not ordered. This random structure is evidenced by the fact that allophanes are amorphous to x-rays and that they have smooth dehydration curves.

These substances were named "allophane" by Stromeyer and Hausmann in 1816 for their changing appearance. Allophanes are highly dispersed mixtures of silica gel and hydrous aluminum oxide or hydrous aluminum silicate. Because of their dispersion, allophanes have a high apparent cation exchange capacity, around 70 milliequivalents/100 g. They actually absorb both cations and anions. The anion absorptivity is high due to the great specific surface and the high aluminum content. Consequently allophane fixes great quantities of phosphate.

Allophane gives a stable porous structure to soils, resulting in high infiltration rates and much leaching. For this reason such soils are frequently infertile.

Oxides and hydroxides of aluminum, iron, and silicon are included in the generic term "clay" and perform a similar function in the soil as the other clays. The majority of these clay-sized particles are electrically neutral and consequently have little or no cation exchange capacity. Some are colloidal gels that react with organic compounds. Frequently they coat the aluminosilicate clays, thus reducing the surface activity of the clay. These neutral sesquioxides and hydroxides occur probably to some extent in all soils but form a large portion of the clay fraction of the latosols.

THE ORIGIN AND OCCURRENCE OF CLAYS

In the classification of clays it seems to be implied that a clay type is something very specific with a crystal lattice and a chemical composition quite apart from all other clays. This may be true for the prototype of the clay, but by and large there are gradations from one type to the other. The reason for this is the origin of clays.

Layer-lattice aluminosilicate clays are (1) formed in the soil as a result of the reaction between silicon and aluminum ions in very dilute solutions, or (2) by the weathering of minerals that form the parent material of soils, or (3) inherited from rocks, usually sedimentary, that already contain clay minerals. Which of these processes is responsible for the origin of the greater amount of clay is a problem. The micas, and to a lesser extent the feldspars, give rise to clays by weathering. In the case of the micas, potassium is removed by percolating waters and the original structure of the mineral remains. Ferrous iron in the lattice may be oxidized, thereby lowering the negative charge of the clay.

The presence of divalent cations, especially magnesium and iron, in the minerals and in the soil solution contributes to the formation of clays of the montmorillonite group. Kaolinitic clays, on the other hand, develop in acid environments of good aeration and strong leaching. This is the reason that tropical soils, exposed to severe weathering and percolation, are usually rich in kaolinite. Montmorillonitic clays are found predominantly in situations of incomplete drainage, especially if the parent material has been well supplied with bases, particularly magnesium.

The clays of the hydrous mica-illite-vermiculite group are predominant in relatively young soils rich in mica.

Allophane, the main constituent of noncrystalline silicate clays, occurs in soils derived from volcanic deposits. They are largely primary minerals. Other noncrystalline clays, such as the oxides of iron, aluminum, and silicon, on the other hand, are largely formed in the soil and can therefore be considered as secondary clays.

THE SIGNIFICANCE OF CLAYS IN SOILS

The presence of clay in soils is of outstanding importance to their physical, chemical, and biological behavior. The great specific surface of clay makes it the active ingredient—along with humus—in the soil. Sand and silt have small specific surfaces and practically no chemical and physical activity. They form the skeleton of the soil. The cohesiveness of clay particles that results from their smallness and their planar surfaces and the ability of some of the clays to swell and to shrink accounts for the important physical role of clay in the soil. Clays act as cement for sand and silt, binding these ingredients together.

Shrinking upon drying separates these soil masses, causing aggregated structure. Below a certain minimum concentration, shrinking of clay does not cause shrinking of soil because the clay can expand and contract within the interstices formed by sand and silt without affecting the positions of these particles. The percentage of clay needed to cause this shrinking of the soil depends on the type of clay, its exchangeable cations, and the proportion of sand and silt in the soil. For surface soils of the

American Middle West 12 to 14 percent is the minimum clay content that will bring about this phenomenon.

The importance of clay on the physical characteristics of soil cannot be overemphasized. Silt without clay has very small pores that are only slowly penetrated by water. Silt is also highly erodible. The same is true of a mixture of sand and silt without clay. Sand without clay has only negligible water-retention power. It should be kept in mind that presence of a fair amount of organic matter would alter these situations.

The most important chemical property of clays is their cation exchange capacity, which to a large extent is the basis of plant nutrition. Some of the three-layer clays, especially montmorillonite and illite, have the ability to fix potassium ions in the interlayer spaces, sometimes with such an energy that the potassium is not released until the clay is dissolved. If there is a generous amount of humus in the soil, potassium fixation is less severe. Two-layer clays do not fix potassium.

All clays fix phosphate, but to very different degrees. The larger the specific surface and the higher the aluminum and iron content, the greater is the fixation. For these reasons fixation increases in the order kaolinite, montmorillonite, allophane.

Clays also contribute to the nutrition of plants by the release of ions upon destruction of the clay crystals by weathering.

The biologic effects of clays are indirect, resulting from the physical and chemical changes just discussed.

METHODS OF IDENTIFYING AND STUDYING CLAY

To identify clays an accurate knowledge of their characteristics and properties is essential. These include the structure of the crystal lattice; the ions making up the lattice; the exchangeable ions within and on the surface of the clay; the size distribution of the clay and its amount present in a given soil.

OPTICAL METHODS

Direct observation of the scintillation of particles in an agitated water suspension indicates the presence of clay. The resolving power of the ordinary optical microscope is about 0.3 micron and permits only the observation of the noncolloidal fraction of the clay and the larger colloidal fraction.

The reason that fine clay appears amorphous under the microscope is that the wavelength of visible light is greater than the truly colloidal clay particles (violet—0.4 micron, red—0.7 micron; this is the same as 4000 and 7000 Å). The reflection of a light beam from the planar surfaces of the clay crystals dispersed in water (Tyndal effect) can be observed by means of an ultramicroscope.

THE ELECTRON MICROSCOPE

About 1935 the electron microscope was developed for the study of matter too minute to distinguish under the most powerful optical microscope. The principle used in the electron microscope is to produce a shadow of the object on a fluorescent screen or a photographic plate by use of a directed beam of electrons. The maximum resolving power of the electron microscope is about 0.4 millimicrons (4 Å). This is 1,000 times as strong as that of the best optical microscopes.

For the study of clay with the electron microscope, the clay crystals must be oriented. This can be accomplished by drying them on a collodion film. The shape, area, and even the thickness of the crystals can be determined. Making replicas of carbon and shadow-casting with heavy metals prior to examination are other useful techniques of studying clays with the electron microscope.

More recently a scanning electron microscope has been developed that permits direct observation of colloidal clay particles (Gray, 1967). This instrument has a great depth of focus and therefore produces an almost three-dimensional effect. Its resolution reaches 0.02 micron. It promises to be of considerable value in the study of the interaction of clay particles with each other and with organic and microbial agencies in the soil.

X-RAY DIFFRACTION

The use of x-ray diffraction permits the study of the spacing of the individual layers in the clay. X-rays are electromagnetic rays of very short wavelength. They were discovered in 1895 by Wilhelm K. Röntgen. They are emitted during a reaction of fast-moving electrons with the inner electrons of metals. The wavelengths of x-rays vary from 0.02 to 10 Å. The shorter the wavelength, the greater is its frequency and the greater is the penetrating power of the x-rays.

X-rays used in clay studies have wavelengths between 1 and 2 Å, depending on the "target metal" used to make the x-rays of uniform wavelength. Frequently copper is used for this purpose. The resulting x-rays have a wavelength of 1.54 Å.

The principle of the x-ray diffraction technique for studying clays is to allow x-rays to strike the clay particles. The x-rays are able to penetrate several atomic layers of the clay lattice and are diffracted by the individual sheets of ions. By measuring the angle at which the reflections from the successive sheets are reinforced it is possible to calculate

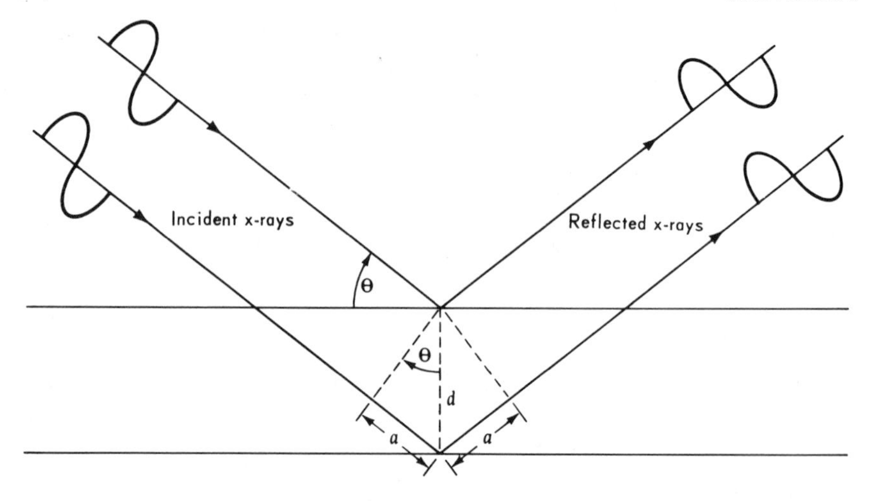

Fig. 4-5 Bragg's law: In order to reinforce each other, the reflected x-rays must be in the same phase. This can happen only if the distance $2a$ is equal to a whole number of wavelengths. Since $a = d \sin \theta$ and both the wavelength of the x-rays and their angle of incidence are known, d, the distance between atomic planes, can be calculated.

the distance between the plates (Fig. 4-5). These spacings are characteristic for the different groups of clay crystals and help to identify them. The distance between the lattice gratings can be calculated using the principle formulated by Bragg (1937) in the following equation:

$$d = \frac{n\lambda}{2 \sin \theta}$$

where d = distance between the lattice gratings
λ = wavelength of the x-rays used
θ = angle of diffraction

Bragg's law shows that a crystal will reflect a beam of x-rays with maximum intensity if n is a whole number. That means radiations reinforce each other, if they are in the same phase.

To prepare a soil or clay sample for diffraction it is dispersed in water using sodium hexametaphosphate as a dispersing agent. The <0.2 micron fraction is separated by centrifugation. For identification of the types of clay, various treatments are needed. Some examples must suffice to explain the reasons for using these treatments.

The typical spacings for clays are approximately 7, 10, 14, and 17 Å. Vermiculite, chlorite, and montmorillonite show 14-Å spacings when prepared with water. When prepared with glycerol, the montmorillonite spacing changes to 17 or 18 Å, while the spacing of the two other clays remains at 14 Å. To distinguish these, they can be heated to 550°C. Chlorite will retain its spacing at 14 Å; vermiculite collapses to about 10 Å.

In x-ray studies kaolinite shows a 7-Å peak. This can be confused with the second-order peak of the 14-Å clays. When kaolinite is heated to 550°C, it is destroyed and loses its x-ray diffraction pattern.

Generally about 2 ml of a 1 or 2 percent suspension of clay in water is placed on a microscope slide for x-ray studies and allowed to dry. This results in essentially oriented clay particles. In some cases random direction of the clay plates is preferred. For this purpose the clay is dried from benzene and the dried clay is again mixed. The possibility of using x-ray diffraction for the study of crystals was discovered in 1912 by Laue, Friedrich, and Knipping. The first use of this method in clay research was made by A. Hadding in 1923 and by F. Rinne in 1924. X-ray diffraction is the most important method of studying and identifying clays.

INFRARED SPECTROSCOPY

The bonds between some atoms and ions vibrate with frequencies which are within the range of infrared radiation. When an infrared beam strikes such an atom pair, energy corresponding to the natural vibrational frequency of the group is absorbed and causes an increase in the amplitude of vibration. Consequently a decrease in the beam temperature occurs and is monitored by a detecting thermocouple. As the wavelength of the infrared radiation is slowly increased from 2.5 to 20 microns (or from a frequency of 4,000 to 500 cm^{-1}), absorption of infrared radiation of specific wavelengths indicates the presence of particular atom pairs (or ion pairs). The wavelengths of absorption of infrared radiation for given ions are determined by using pure chemicals containing these ions. Measuring the vibrational energy levels by infrared spectroscopy permits the determination of the groups present in the clay structures. The orientation of the bond axes of structural OH groups is different for dioctahedral and trioctahedral minerals, and these can be distinguished by their infrared absorption spectra. The infrared spectrum is characteristic for the various types of clay, thus making an identification possible. This method is also useful to study complex formation between clay and organic matter.

One of the great advantages of infrared spectroscopy is that clays which are amorphous to x-rays can be studied.

CHEMICAL METHODS

The chemical composition of the clay crystal as such is considered here, not the exchangeable ions that may be associated with it. For that reason it is necessary to replace the exchangeable cations with hydrogen. This is done by dialysis or with an exchange resin. Colloidal and other impurities have to be removed.

Organic matter can be oxidized with H_2O_2 or with sodium hypochlorite. Iron oxide is dissolved in oxalic acid or with sodium sulfide. The silica-alumina ratio and the silica-sesquioxide ratio together with the amounts of cations present, particularly potassium, can give an indication of the type of clay. High silica-sesquioxide ratios point to montmorillonitic clays; low silica-sesquioxide ratios are generally associated with kaolinitic clays. The presence of relatively high or relatively small amounts of certain elements are characteristic of certain clays. For instance, muscovite, biotite, hydrous mica, and illite are high in potassium (6 to 8 percent K). Vermiculite has around 15 percent magnesium (Mg), while illite, montmorillonite, and biotite are also fairly rich in this element (around 3 percent Mg). Nontronite is lower in aluminum (less than 1 percent Al) and higher in iron than most clays.

Although quantitative determinations of the chemical composition of clays have some value in verifying the identity of a clay, by themselves they cannot identify a clay. This is because of the similarity in composition between various clay types and because of the difficulty of removing interfering impurities. Also isomorphic substitutions complicate the interpretation of the results.

The exchange capacity for cations can be determined and used as an indication of the types of clay.

THERMAL METHODS

Thermogravimetric analysis The water in various clays is given off at different temperatures. It is necessary to distinguish clearly between constitutional water (OH groups) and interlayer or adsorbed water (molecular H_2O). It is possible to heat clay samples and to determine their moisture losses with increasing temperature. The resulting curves for the main types of clay are quite distinct and thus can be used in the identification of the major types of clay. Kaolinites give off most of the hydroxyl water around 550°C, while the interlayer water of montmorillonites is given off between 150 and 250°C and more hydroxyl water at temperatures above 550°C. The exact temperature level at which water is given off by clays depends on the rate of heating. The slower the heating, the lower the temperature at which water is given off.

Differential thermal analysis In this method use is made of the fact that the driving off of water from the clay crystals and the driving off of the water of hydration of the exchangeable ions are endothermic reactions (reactions that absorb heat). This causes the temperature of a clay sample to fall below the temperature of a similarly treated inert material. Other endothermic and exothermic reactions occur as a consequence of rearrangements of atoms in the crystal lattice at temperatures specific for the various types of clay. A sample of clay is heated simultaneously and with the same amount of heat energy as some inert material (e.g., calcined alumina) of similar specific heat and heat conductivity. Temperature differences between the clay and the inert material are recorded. These are plotted against temperature and give curves that are specific for each type of clay. This method is very useful in identifying clays. From the intensity of the endothermic reactions the proportion of the various clays in a mixture can be estimated. This method is particularly useful to identify the two-layer silicates, especially kaolinite and gibbsite.

Heat of wetting Heat of wetting is an exothermic process. When dry clay absorbs water, heat is liberated. The amount of this heat depends on the nature of the clay, its specific surface, and the exchangeable ions present. The determination of the heat of wetting is seldom used in the identification of clays because better methods exist.

IMBIBOMETRY

Recently a qualitative method of clay identification has been suggested (Konta, 1963) that is based on the rate of intake of certain liquids in compressed samples of clay. The rate of intake, the amount of swelling, and the area moistened by a drop of a definite size are indicative of the specific surface of the clay. So far, this "imbibometry" method has not been generally adopted.

REFERENCES

Black, C. A. (ed.): Physical and Mineralogical Properties, Including Statistics of Measurement and Sampling, "Methods of Soil Analysis," pt. 1, Am. Soc. of Agron., Inc., 1965.

Bragg, W. L.: "Atomic Structure of Minerals," Cornell University Press, Ithaca, N.Y., 1937.

Gray, T. R. G.: Stereoscan Electron Microscopy of Soil Micro-organisms, *Science,* vol. 155, pp. 1668–1670, 1967.

Grim, R. E.: "Clay Mineralogy," McGraw-Hill Book Company, New York, 1968.

Konta, J.: Identification of Clay Minerals, *Proc. 10th Natl. Conf. Clays and Clay Minerals,* vol. 10, 1963.

Marshall, C. E.: "The Physical Chemistry and Mineralogy of Soils, I, Soil Materials," p. 388, John Wiley & Sons, Inc., New York, 1964.

Rich, C. I., and G. W. Kunze (eds.): "Soil Clay Mineralogy," p. 330, The University of North Carolina Press, Chapel Hill, N.C., 1964.

5 SOIL STRUCTURE

ARCHITECTURE OF SOILS AND THE MEANING OF SOIL CONSISTENCE AND SOIL STRUCTURE

The nature of soil is determined not only by the individual properties and ingredients but largely by the relationships of these factors to each other. The study of the relationships of the various physical properties and ingredients to each other might be called the *architecture* of soils.

Some of these relationships are rather obvious and are only enumerated here. Others are extremely important, such as consistence and structure. The latter particularly has been recognized as so fundamental to any understanding of soil behavior that the largest part of this chapter is devoted to it.

Items of the architecture of soils are

1. *Relative location:*
   a. Physiographic location, microclimatic influence of upland, valley, depression, ridge, and of steepness, configuration, and aspect of slope. Availability of precipitation water, opportunity for evaporation and condensation, temperature situation, likelihood of freezing.
   b. Location with respect to groundwater.
   c. Location with respect to other horizons.
2. *Vegetation:* Type and amount of vegetation affect the temperature, moisture, and structure of the soil. Decayed roots leave open channels.

3. *Animal activity:* The burrows of animals supply a system of communication for water and air in the soil. Animal droppings represent the largest part of well-aggregated surface soil.
4. *Consistence:* The holding together of soil particles.
5. *Structure:* The arrangement of individual soil particles with respect to each other.

SOIL CONSISTENCE

DEFINITION AND PHENOMENA OF CONSISTENCE

> Soil consistence comprises the attributes of soil material that are expressed in its degree and kind of cohesion and adhesion or in its resistance to deformation or rupture. (Templin *et al.*, 1947)

The phenomena of soil consistence are friability, plasticity, stickiness, and resistance to compression and shear, sometimes referred to as soil strength or bearing strength.

FORCES CAUSING CONSISTENCE

Two main forces are responsible for soil consistence: molecular attraction (cohesion) and surface tension (adhesion). In addition to these, other factors contribute to consistence, e.g., organic compounds, iron and aluminum oxides and hydroxides, and calcium carbonate. A special case of soil consistence is brought about by freezing of the soil water.

Cohesion Molecular attraction is brought about by the surface charges of the clay particles, by the broken bonds at the edges of the plates, and by the attraction from particle to particle (van der Waals' forces). Consistence as a result of molecular attraction can consequently be large only if the soil particles lie closely together and have relatively large surface areas in common. This is particularly the case with oriented colloidal clay. The effect of molecular attraction is greatest in dry soil and sharply decreases as water enters between the particles, thus keeping them apart. Molecular attraction (cohesion) is most effective when the individual particles (especially the clay) are oriented so that they lie close to each other (thixotropy) (Fig. 5-1).

Adhesion Surface tension (film forces) has been discussed previously. Its effect on holding soil particles together depends on the presence of both water and air.

While surface tension per unit area of contact is greatest where the meniscus is curved the most, as it is in soils of limited moisture content, the total area of contact is fairly small under these conditions. As the soil moisture increases, the area of contact increases while the surface

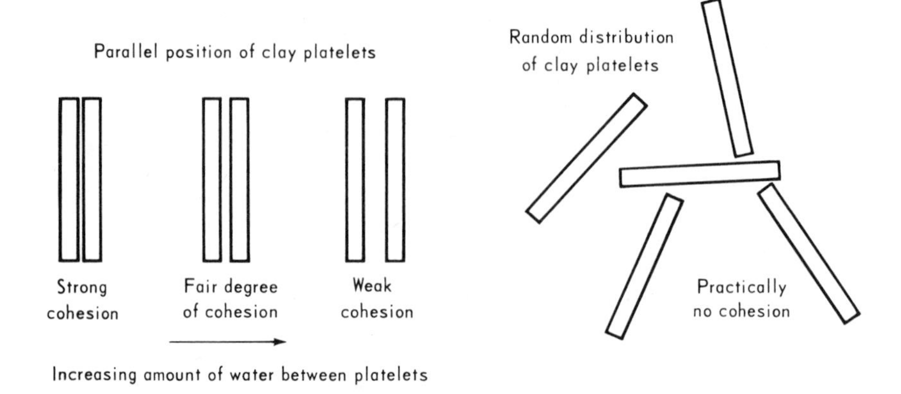

Fig. 5-1 Water content and relative position of clay particles affect cohesion (schematic diagram).

tension per unit area decreases only little. The result is that the consistence increases. At a certain point in the wet range, however, surface tension per unit area has become so small that even the increased area of contact fails to compensate for it. Consequently, consistence decreases. Eventually, when the soil is fully saturated with water, surface tension ceases to exist (Fig. 5-2).

Since consistence is a result of both molecular attraction and surface tension, it has two maxima and two minima. Consistence is large at the extreme end of the dry range, due to cohesion, and again in about the middle of the wet range, due to adhesion. It is small in the moist range and when the soil is saturated (Fig. 5-3).

These relationships are true only for soils with sufficient clay contents to permit molecular attraction to become effective. Molecular attraction of silt and sand is insignificant. As soon as the consistence of a dry soil has been destroyed and the soil particles separated it can only be re-established by rewetting the soil, thus orienting its particles again.

In addition to the amount of water and clay in the soil, other factors affect its consistence:

The type of clay: Montmorillonitic clay causes more consistence than kaolinitic clay.

Texture: Cohesiveness increases with decreasing particle size.

Organic matter: It causes more cohesion than sand and silt, but less than clay.

Structure: A puddled soil has more cohesion than a well-aggregated one, because it has more areas of contact between the individual particles. Puddling is orienting the clay particles so that they lie parallel to

	Moisture condition	Meniscus		Adhesion
		Curvature	Contact area	
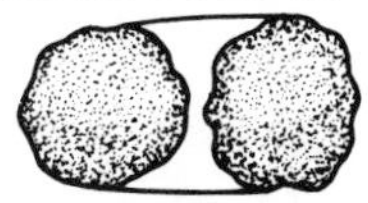	Wet	Not curved	Large	Almost none
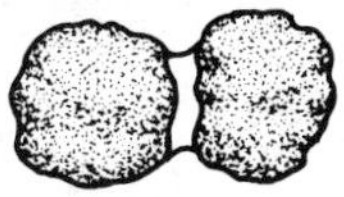	Fairly wet	Curved	Medium	High
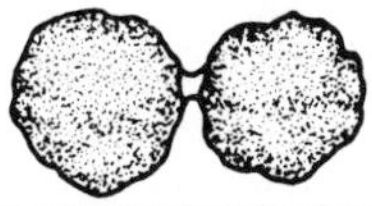	Slightly moist	Sharply curved	Small	Low
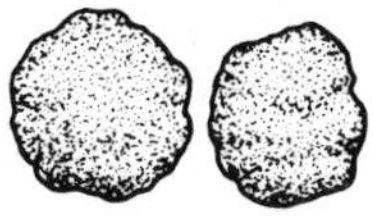	Dry	None	None	None

Fig. 5-2 Curvature and contact area of water meniscus affect the consistence of soil.

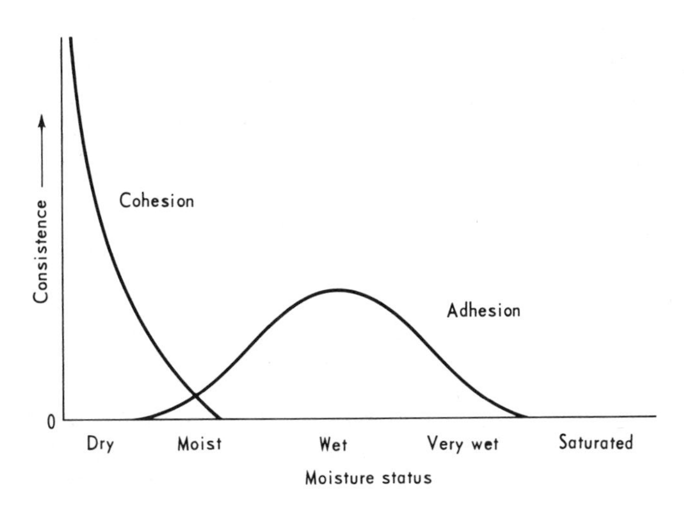

Fig. 5-3 The effect of moisture on the two main components of soil consistence (schematic diagram).

each other and filling the large pores with soil particles, thus decreasing pore space. Upon drying the consistence is greatly increased because of the increase of surface contacts. Also, a dispersed colloid causes more consistence than a flocculated colloid.

FORMS OF SOIL CONSISTENCE

The different forms of consistence, largely resulting from differences in soil-moisture contents (or energy levels) are

1. Dry soil, hard or harsh consistence; soil particles when once apart do not stick together again. Soil consistence in dry soils depends on the amount of surface contacts per unit volume.
2. Moist soil, friable consistence; soil sticks together only lightly, and it is soft.
3. Wet soil, plastic consistence; soil can be molded without loosing its coherence.
4. Very wet or saturated soil, viscous consistence; soil may either flow under pressure or merely by gravity. Soil adheres to objects, and it is sticky.

Table 5-1 is presented to help to clarify the relationships between soil moisture, soil consistence, and some associated properties. The soil-moisture-tension limits are merely indicative. They are not based on actual determinations. They do not include the osmotic component of the total moisture stress.

PLASTICITY

Plasticity is one range of consistence. It is the capacity of being molded. This implies a change of shape without cracking, when the soil mass is subjected to a deforming stress. Since wet clay is the only plastic material in soil, soils with less than 14 to 16 percent clay do not exhibit plasticity in any moisture range. The amount of clay necessary to make a soil plastic depends on the type of clay, the relative amounts of silt and sand present, and the percentage of organic matter. The higher the clay content, the wider the moisture range over which a soil is plastic. This range is called the *plastic range* or the *plasticity number*. Consequently, a wide plastic range indicates a soil that is hard to handle because of much sticky clay. The plasticity number is expressed in percentage of water between the liquid and the friability limits of the plastic range. The lower plastic limit or the friability limit is the boundary between friable and plastic consistence. The upper plastic limit or the liquid limit is the boundary between the plastic and the liquid consistence. Atterberg (1911, 1912), a Swedish soil scientist, devised methods to determine these limits, which are still widely used by soil engineers.

Table 5-1 Effect of soil moisture on the consistence of soils of medium to high clay contents

Soil moisture	*Dry*	*Moist*	*Wet* *Fairly wet*	 *Very wet*	*Saturated*
	Oven dry pF 7.0	Hygroscopic coefficient 4.5	Friability limit 2.8		Zero tension 0
Consistence forms	Hard, harsh	Friable, soft	Plastic, sticky		Viscous, soil flows by gravity
Relative degree of consistence	Very high	Low	High		Very low
Agency mainly responsible for consistence	Molecular attraction (cohesion)		Surface tension (adhesion)		
Bearing strength	High	Fairly high	Low	Very low	Practically none
Ease of tillage	Heavy draft	Light draft	Heavy draft, implements tend to sink and slip	Draft lighter, but traction low, imple– ments frequently bog down	Ordinary tillage impossible
Result of tillage	Soil forms clods and dust	Soil crumbles, optimum effect on structure	Soil puddles	Soil runs together	
Graphic represen– tation of relative degree of consistence					

The lower plastic limit is the minimum moisture condition at which the soil can be puddled. The bearing strength rises sharply as a soil dries beyond the lower plastic limit. Organic matter raises the amount of water needed to satisfy the clay and to make the soil plastic. It consequently raises both the liquid limit and the friability limit but leaves the plasticity number essentially unaffected.

MEASURING CONSISTENCE

Some of the methods devised to measure soil consistence are designed for natural conditions, while others measure the consistence of artificially treated and molded soils. The latter are of little direct value for agronomists. The methods include the determination of the resistance to penetration of a special probe (penetrometer), the draft requirements of a plow or similar implement with a dynamometer or a strain gage, the Atterberg plasticity constants, the crushing resistance, the modulus of rupture, the tensile strength of a plastic soil thread, the bearing strength, the compression resistance. Methods of measuring various phases of consistence have been described by Richards (1953), Sowers (1965), Felt (1965), and Davidson (1965).

SOIL STRUCTURE

DEFINITION AND IMPORTANCE OF STRUCTURE

The arrangement of the individual soil particles with respect to each other into a pattern is called *soil structure.* Since the pore spaces are as important in a soil as the solid particles, soil structure may also be defined as the arrangement of small, medium, and large soil pores into a structural pattern. Such a definition points out one of the most important primary effects of soil structure.

Soil structure as such is not a plant-growth factor, but it influences practically all plant-growth factors. Water supply, aeration, availability of plant nutrients, microbial activity, root penetration, all these and more factors are affected by soil structure. Consequently, poor soil structure may be an indirect factor limiting plant growth. On the other hand, good soil structure permits the plant-growth factors to function at optimum efficiency.

CLASSIFICATION OF SOIL STRUCTURE

Many soil scientists have attempted to classify soil structure, but to date no universally accepted classification exists, probably because no satisfactory method of measuring soil structure has been devised. Some of the aspects of structure can well be determined but as yet there is no way of describing structure as such quantitatively.

The following classification is adapted from various sources. It is greatly simplified and omits many subtypes.

Classification according to shape

1. *Simple structure:* Natural cleavage planes absent or indistinct.
   a. *Single-grain structure:* This occurs normally only in sands and silts of low organic matter content. In sands single-grain structure allows for aeration and a maximum of capillary water movement. In all other soils it is undesirable because it results in complete absence of large pores needed for satisfactory aeration.
   b. *Massive structure:* Massive structure is similar to single-grain structure, except that it is coherent. Examples of massive structure are dense soil crusts, plow pans, and fragipans.
2. *Compound structure:* Natural cleavage planes distinct. The shape of the individual peds of soil of compound structure can be described according to the relative lengths of the vertical and horizontal axes and by the contours of their edges. The forms of peds are:
   a. Cubelike structure, vertical and horizontal axes of similar length.
   b. Columnar structure, prismatic, vertical axes longer than horizontal axes. Vertical cleavage planes predominant.
   c. Platy structure, horizontal axes longer than vertical axes. Horizontal cleavage planes predominant.

   The contours of the edges of the peds are:
   a. Angular, corners and edges are sharp and distinct.
   b. Subangular, corners are rounded, edges are sharp.
   c. Granular, both corners and edges are rounded.

   While the sharp edges are generally the result of physical forces, such as wetting and drying or freezing and thawing, the rounded surfaces owe their origin to organic influences.

Classification according to hardness of aggregates The hardness of the aggregates is influenced by the moisture content, the amount of clay, the type of clay, the nature of the adsorbed cations, and the organic matter content. High moisture content, kaolinitic clay, divalent cations, and high organic matter content make for relative softness of the aggregates.

Since moisture content is of outstanding effect on the hardness of soil, the hardness of aggregates should be compared at the same moisture tension.

Classification according to size of aggregates Since sand-sized aggregates are more favorable for plant growth than very small and very large ones, size of aggregates is a valuable criterion of structure. Note that a soil made up exclusively of silt-size particles or aggregates cannot be drained by gravity, since the pores are too small. It requires a tension of ½ atmosphere to empty a pore of 0.006 mm in diameter. Most pores in pure silt are smaller than this.

Classification according to stability Aggregates vary greatly according to the way they stand up under the impact of raindrops or under submergence. This stability depends upon the clay content, the extent of flocculation, the organic-inorganic linkages, the microbial glue of the aggregates, and the presence of mineral cementing materials, such as iron and aluminum oxides.

Classification according to pore sizes Since plant roots and microbes live in the pore spaces, classification of soil structure according to pore sizes may be quite logical. The percentage distribution of pore space of various sizes can be determined by removal of water from a saturated soil sample. The resultant data when plotted as a soil-moisture-desorption curve give a good picture of the structural pattern of a soil.

GENESIS OF COMPOUND SOIL STRUCTURE

Factors inducing aggregation *Clay and exchangeable ions:* Generally clay assists in aggregating soil by acting as a cement and also by its ability of swelling and shrinking with moisture changes. Thin films of clay cover small aggregates of particles of certain soils and serve to hold these aggregates together. Such films are called *clay skins* and help to distinguish between soil groups. Leaching out of clay has proceeded to such an extent in latosols that clay skins are absent. Because of their small zeta potential kaolinitic clays flocculate more readily than montmorillonitic clays. Exchangeable ions have also a pronounced effect on the flocculation of clays.

Generally calcium, magnesium, and potassium have a flocculating effect on clay, while hydrogen and especially sodium have a peptizing effect.

← flocculating, peptizing →

Ca, Mg, K H, Na

It should be remembered that flocculation and aggregation are not at all the same thing. Flocculation is an electrokinetic phenomenon that we usually study in aqueous suspensions.

The flocculated clay masses are generally not larger than silt size. This means that flocculated clay would be a very unfavorable environment for plant roots, unless it is further aggregated. It so happens that calcium and to a lesser extent magnesium induce flocculation of clay and at the same time are important for the nutrition and for a favorable reaction for the growth of plant roots and soil microbes. A soil devoid of organic matter does not become readily aggregated by the addition of lime.

Sodium on the other hand tends to peptize (deflocculate) clay and to create a reaction in which several of the plant nutrients become unavailable, consequently depressing the growth of roots and microbes. Therefore, a poor structure results.

Inorganic cements: Sesquioxides form irreversible and slowly reversible colloids. These help to form water-stable aggregates. This effect is particularly noticeable in the latosolic soils of the tropics. In spite of high contents of acid clay these soils are generally well aggregated. The smallest peds are of fine-sand size. This results in pores large enough to allow rapid percolation of excess water. Calcium carbonate precipitating out around soil particles also acts as a cement. Salts tend to enhance flocculation of clay, even of sodium-saturated clays.

Plants and plant residues: Plants help to aggregate soil particles in a variety of ways. The most important one probably is the excretion of gelatinous organic compounds from the roots which serve as links between inorganic substances. Root pressure, respirational carbon dioxide, and excreted minerals may also be of value. Root hairs make the soil particles cling together. Dehydration of the soil by the roots causes strains in the soil due to shrinkage that may result in cracks and subsequently in aggregate formation.

Plant tops and residues keep the soil shaded and protect it from extreme and sudden temperature and moisture changes and from raindrop impact.

Plant residues, both roots and tops, serve as food supply for microbes, the prime aggregate builders. Continuously new and biologically active organic matter is required for this purpose. Plant residues also serve as insulating material between soil cracks.

Organic compounds and cements: Organic compounds are effective in inducing aggregation and in stabilizing soil structure. They form irreversible or slowly reversible colloids that serve as cements. Not all organic matter stabilizes soil. Some of the simpler compounds, such as the sugars, are ineffective before they are changed to microbial tissue and decomposition products. Fats, waxes, lignins, proteins, resins, and some other organic compounds have a direct stabilizing effect.

The cations associated with humus are of great importance for soil structure. Only Ca humus is flocculated; Mg, K, H, and Na humus is peptized and does not aid in aggregation. Maybe the most important factor in aggregating soil particles is the complex formation of organic compounds in linkages with metallic ions (chelation).

Organic colloids serve also as insulating cover of low consistence between cracks. Organic matter may be washed into them or it may form in place.

A number of organic compounds have been used as so-called soil conditioners. These are mostly long carbon-chain compounds that attach themselves to the exchange sites of the clay and thus link together many clay particles.

Microbes: Flocculation of clay is in general not sufficient for aggregate formation, especially not for the formation of stable aggregates. Algae, fungi, actinomyces, and bacteria form living matter that will keep soil particles together more effectively than any exchangeable ion could. Such a coating corresponds to the plant cover that protects the general soil surface from erosion.

Plant residues alone do not cause aggregation of soil particles. Sekera and Brunner (1942) treated soil with alfalfa meal (½ percent) that was poisoned with HgCl. No structure improvement resulted, while without poisoning the aggregation increased considerably on addition of alfalfa meal. They also found that CO_2 produced in the decay was not the cause of aggregation (by increasing the Ca^{++} and HCO_3^- concentrations) as they tried CO_2 additions to the soil without success.

Animals: Some workers claim that the majority of the true humus represents the metabolic products of small animals. These include the earthworms, spiders, mites, springtails, nematodes, insects, and many others. Their combined effect on structure formation is probably very great.

It has been estimated that the earthworm produces about its own weight in water-stable aggregates in one day, if conditions are favorable. It also prepares channels that are useful for infiltration, drainage, and aeration. Generally, the greater the earthworm population of a soil, the better is its aggregation. This may not necessarily be cause and effect; it may be that the same soil conditions are favorable for both earthworms and aggregation.

Water: Water participates in soil-structure formation in a variety of ways:

1. Swelling and shrinking of colloids, brought about by wetting and drying, develops planes of weakness by causing strains and pressures in the soil body. The cleavage planes due to drying are predominantly in a vertical direction. Very dry seasons cause aggregate formation in the subsoil. Columnar structure is the result of drying the soil profile beyond the freezing depth.

2. Surface tension, resulting from the presence of water and air surfaces, keeps soil particles together and orients them. (This was discussed under "Cohesion.") On the other hand, surface tension forces water into spaces previously filled by air and thus causes minor explosions when the air escapes. In this way soil of massive structure is broken up into smaller aggregates. Although this is generally desirable, the same process can decrease the size of aggregates too much.
3. The effect of ice formation upon soil structure depends on the moisture content of the soil, the pore-size distribution, and the velocity of freezing. Little effect can be expected upon soil structure if the soil is nearly dry.

Slow cooling and freezing of a moist soil is beneficial for soil structure; water begins to freeze in the larger cavities where it is under the least tension. In the vicinity of the ice crystals the soil gets drier and water will move toward these drier regions due to the tension gradient and diffusion. More water will freeze around the original ice crystals and this process will continue until all "free" water is frozen, unless temperature changes have occurred.

The effect of this process is that through dehydration particles are held together by strong surface-tension forces and by concentrating the electrolytes present. The pressure of the large ice crystals increases the pores and presses aggregates tighter together. Slow freezing holds the small aggregates together, but breaks up the bigger soil masses by cracking them at the natural "surfaces of weakness." Since freezing begins at the soil surface and water moves up toward the frozen zone, the ice formation is normally parallel to the surface. Therefore cleavage planes due to freezing are mostly in horizontal direction.

Fast cooling and freezing of a wet soil causes ice-crystal formation throughout the soil mass. This results in breaking apart of aggregates, i.e., in a dispersion of the soil, which is very undesirable. It may have the advantage of breaking down large clods, but the aggregates thus formed are stable only when the soil is naturally well aggregated. Otherwise, the frost-induced structure is evanescent.

Frequent freezing and thawing of the soil surface result in an accumulation of water in the surface to such an extent that the soil may become thoroughly dispersed. As part of the water reaches the surface soil by distillation, it is low in electrolytes and the zeta potential of the clay will increase due to dilution of the soil solution causing dispersion.

The great effect texture has on water movement in soil during freezing and on the resulting structure is illustrated in Figs. 5-4, 5-5,

Fig. 5-4 Homogeneous structure of a frozen sandy soil. The dark area in the center represents a thin ice layer that occurred during the experiment. (*Walter Czeratzki.*)

Fig. 5-5 Horizontal structure of a frozen loam. Dark horizontal lines are solid ice. (*Walter Czeratzki.*)

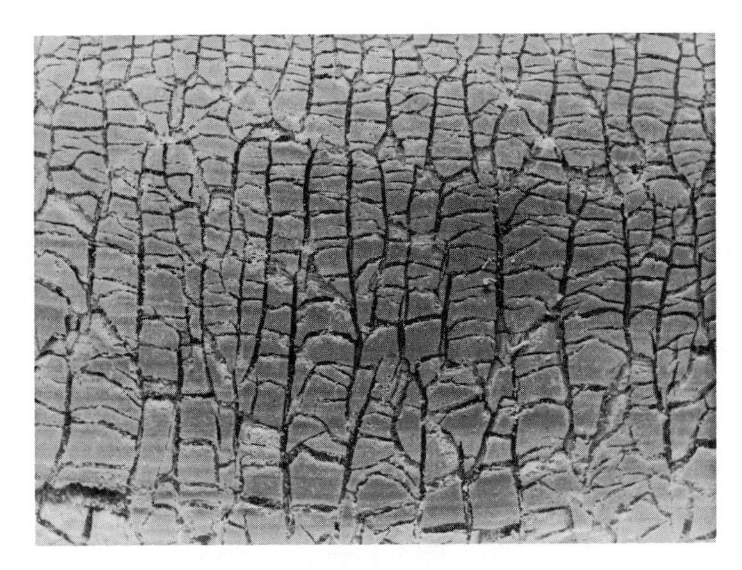

Fig. 5-6 Polyhedric structure of a frozen calcium montmorillonite. Horizontal and vertical cracks are filled with solid ice. (*Walter Czeratzki.*)

and 5-6. In all cases the soils were amply supplied with water before being frozen. The sandy soil (Fig. 5-4) shows little effect of freezing on structure. The pores are so large that water is under very little tension and freezes as soon as the temperature has dropped to 0°C. Since most pores were not completely filled with water before freezing, the 9 percent expansion of water can be accommodated without substantial rearrangement of the solid particles. The result is a very slight loosening of the structure.

In the loam the situation is quite different (Fig. 5-5). The soil has a much greater capillary conductivity. In the fine pores the water is held under high tension, so that its freezing temperatures lie considerably below 0°C. As the 0°C isotherm descends into the soil, water in the large pores freezes. The liquid water in the capillary pores below this line moves up to the ice layer, helping this to increase in thickness. The heat of fusion keeps the soil from cooling down too fast. If the progress of the 0°C isotherm is sufficiently slow, much of the water in the soil matrix below the ice lense is removed in this way and the remaining water is under such high tension (has such a low free energy) that it cannot freeze before the temperature has dropped considerably below 0°C. Farther down a new ice lense is formed as water in a layer of larger pores. The thickness of the ice lenses depends largely upon the velocity of freezing. The slower the freezing, the more time there is for the water to move up to the ice lense before the temperature below it has become so low as to also freeze some of the water in the matrix below the lense, thus stopping further capillary movement. When freezing is fast, ice lenses are only thin, since the temperature in the soil below the lense quickly drops to a point where capillary movement is stopped.

Freezing of water in a clay (Fig. 5-6) has a particularly interesting effect on soil structure. The natural phenomena are similar to those in the loam. The difference is that the pores in the clay are much finer and consequently capillary movement is much slower and the water is held under a much higher tension. Therefore it freezes in the matrix of the clay at a still lower temperature than in the matrix of the loam. In addition, clay shrinks upon removal of water. When clay freezes lense formation occurs in the same way as in the loam. The upward movement of the water to the ice lense results in shrinking of the soil substance and consequently in vertical as well as horizontal separation. The resulting cracks are large enough that water freezes in them at very nearly 0°C. These cracks increase in thickness in proportion to the added water that freezes in them. At the same time the polyhedra of the clay continue to shrink.

To what extent freezing of a soil will result in a laminated structure parallel to the surface or in blocky structure depends on the amount of expanding clay it contains.

4. Water as a prerequisite of microbial and higher plant life. Proper amounts of water throughout the growing period assure maximum biologic activity in the soil, as far as other growth factors permit. Maintenance of reasonable moisture in the soil is one of the prime requisites for genesis of soil structure. This moisture content can vary considerably and still permit very active microbial life. It is probable that a limited variation in soil-moisture tension may even be beneficial in this connection. Besides, the optimum moisture range for various microbial species is not identical.
5. Water as a climatic factor. A few examples will illustrate how the amount and temperature of precipitation have a pronounced effect on soil structure.
   a. *Deserts:* Climates with no or very little rainfall allow little chemical weathering and clay formation and little formation of organic matter and consequently little microbial life. Hence, as all chemical and biological factors of soil genesis exist in minor amounts, aggregation and structure formation are small.
   b. *Chernozems:* In climates where precipitation and evapotranspiration are approximately equal, there is a good deal of clay formation. As calcium is not leached out and organic matter is produced in

fair amounts and is not decomposed too quickly, calcium helps to flocculate soil and organic matter. These conditions bring about a favorable environment for microbes, except for occasional drought periods. Soils of this climatic province are excellently aggregated.

c. *Podzolic soils:* Cool humid climates permit much clay formation and plant growth, but little accumulation of clay and plant nutrients is possible because of extensive leaching. The bases and particularly calcium are leached from the soil. This makes for formation of hydrogen humus which is not stable. Consequently little material exists for structure formation and these soils are generally poorly aggregated. This is particularly true of the A_2 horizon of the podzols. Liming and fertilizing are the first steps to induce a better structure in such soils.

d. *Latosols:* Where a large amount of leaching coincides with high temperatures, the drainage water becomes alkaline because of the very rapid decomposition of organic matter and the release of bases and because warm water contains only very low concentrations of carbonic acid. The result is that silica is leached from the soil and iron and aluminum oxides predominate. These help to form a very stable structure, even though there is only little organic matter present.

Air: Physically air plays a role in soil-structure formation as a necessary component in creating surface tension and in the air explosions due to wetting a soil mass. When water touches an aggregate of dry soil lying on the surface of the ground, as happens in a severe rainstorm, this water is attracted with great force. In this way the air in the center of the aggregate is compressed and tends to escape. Unless large pores serve as avenues of air release, the aggregate breaks up. Chemically air assists in the precipitation of iron and aluminum colloids. The high content of CO_2 in the soil air in equilibrium with that of the soil solution helps bring calcium and other ions into solution which are important links in aggregation. Biologically air is necessary for the respiration of roots and microbes. The importance of air in structure formation and maintenance is so great that an extended period of flooding deteriorates soil structure.

Under anaerobic conditions the organic compounds that serve to cement the soil particles are not formed. Those that exist are rapidly decomposed. Under flooding clay is highly hydrated and does not hold together because of the larger distances between the individual particles, especially if the water is low in electrolytes.

Temperature: The effects of temperature on soil-structure formation are mostly indirect. Physically temperature induces water vapor movement by creating vapor pressure differences. The effects of freezing and thawing have been discussed before. Chemically temperature affects the speed of reactions and consequently the decomposition of organic matter and the weathering of minerals. Biologically temperature is of outstanding effect upon the activity of plants and microbes. Temperature as a climatic factor has much influence upon soil structure. The types of clay formed, their zeta potentials, the exchangeable ions present, the amount of soluble salts, the amount of organic matter, all are influenced by the temperature of the climate.

Pressure: Soil particles, especially clay and organic colloids, when pressed together have the tendency to become oriented and to stay together due to molecular attraction and film forces.

Pressure is created in the soil in a variety of ways; for instance, by wetting and drying causing swelling and shrinking, freezing, root pressures, agricultural implements, and the weight of the overlying soil itself.

Factors stabilizing compound soil structure Creation of a certain type of compound soil structure does not in itself imply that this structure will persist for any extended period of time. Many of the physical and chemical factors of soil-structure formation bring about aggregates that lose their identity when they dry out or when they are hit by raindrops or for other reasons, such as frost or pressure. It is generally the biologic and organic factors that protect soil aggregation. Root exudates, bacterial slimes, and such organic compounds as pectin are of prime importance.

Exchangeable ions and soluble salts contribute to flocculation of clay and organic colloids and therefore to the stability of the aggregates.

FACTORS INDUCING DISPERSION

After the factors inducing aggregation have been discussed it is not necessary to dwell at length on the factors inducing the opposite process. Excessive conditions of many otherwise favorable phenomena bring about dispersion of soil structure, e.g., excessive wetness, dryness, oxidation, reduction, leaching, heat, pressure, and cultivation. Also intensive rainfall, frequent and fast freezing and thawing, and erosion act in the same direction.

THE EFFECT OF TEXTURE ON STRUCTURE

Sandy soils are generally porous, tending to single-grain structure. If sandy soils have a fair content of silt and clay, but not enough for good aggregability, they become massive upon working (puddling). Silts are very tight, have small pores, and are hard to drain and therefore are poorly aerated. Silt loams, especially if they have a fair amount of organic matter, may be well aggregated. Silt loams low in organic matter have unstable structure and crust easily. Soils of more than 12 percent expanding clay may have a structure favorable for plant growth, if they are subjected to freezing and thawing and wetting and drying. This is especially true if they contain sufficient organic matter to help stabilize the aggregates. Soils of clay contents of more than 35 percent are usually massive and unfavorable for plants. An exception to this are the tropical soils which contain principally nonexpanding clay and are aggregated by iron and aluminum oxides.

THE STRUCTURE PROFILE

In most soils there is a change of structure with depth. Generally speaking soils of A_1 horizons have granular structure because they are rich in organic matter and soil flora and fauna, while soils of B horizons have angular structure. A forested podzolic soil may have crumb structure in the A_1 horizon, dispersed soil in the A_2 horizon, block structure in the B horizon, and massive structure in the C horizon. Soils developed under forest cover have much sharper differences between the structures of their individual horizons than soils developed under grass. Also, age of the soils has an effect on the development of structure. Therefore, mature soils have more pronounced structure profiles than young soils. Very old soils, such as latosols and the older planosols, have only gradual horizon changes.

The part of the profile whose structure is perhaps of greatest influence upon plant growth is the top ½ cm. Sometimes soils of otherwise good structure have a dense crust that may impede the movement of air and water and hinder the emergence of seedlings. Another horizon that frequently is less pervious than those above and below is the plow sole. Such sharp differences between adjoining horizons are detrimental to plant growth since they tend to impede water intake and aeration. This condition is sometimes brought about artificially on golf greens, if layers of different texture and organic matter content are placed alternately on the area in question.

DESIRABLE SOIL STRUCTURE

The criterion of soil structure is plant growth. Agriculturally the best structure is the one that results in the greatest crop yields. Structure quality can be expressed in terms of porosity, aggregation, cohesiveness, or permeability for water or air. Probably the most meaningful classification of soil structure is based on porosity, since the chemical and biologic processes occur in the pores.

Porosity Large pores serve aeration and infiltration, medium-sized pores serve water conduction, and small pores serve storage of plant-available

Table 5-2 Soil pore sizes and their functions

Pore diameter, mm	Corresponding matric tension, pF	Pore function classification	Biotic limits
0.00003	5	Hygroscopic surfaces	
0.0002	4.18		
0.0003	4	Storage of plant available water	
0.001			Bacteria
0.003	3		
0.009	2.54	Capillary conduction	Root hairs
0.02			Protozoa and algae
0.03	2		
0.06	1.7	Aeration porosity – Fast drainage	
0.1			Rootlets
0.3	1		
1.0	0.47		

water, while hygroscopic water is held on the surfaces of all soil particles and within the framework of the expanding clays (Table 5-2). What proportion of these pore ranges to each other is most beneficial for plant growth depends on the species grown but particularly also on the climatic conditions and on the height of the soil above groundwater and on the possibility of irrigation. Where much water is available from rain or irrigation, water-storage pores (small pores) are not so important, but aeration pores (large pores) are greatly needed.

In a "dry farming" area of restricted rainfall, storage pores are vital and sufficient large pores are needed to ensure adequate infiltration capacity. In humid climates large, medium, and small pores should occupy about equal volumes.

Table 5-3 gives a schematic picture of pore requirements in different climatic areas; it is not intended to give any quantitative information.

Aggregation The aggregates of a soil should be arranged in such a way as to give the pore distribution described above, and they should be stable enough to retain their identities in spite of rainfall impact or temporary submergence. The vertical axis of the aggregates should be as long as or longer than the horizontal axis. Pronounced horizontal cleavage planes should be absent. Rounded edges of the aggregates result in better pore distribution than angular edges. Aggregates of sand and gravel size are preferred. If such aggregates are water stable, the soil is said to be in good tilth. Optimum size of peds for good plant growth, particularly early in the growing season, lies between 0.5 and 2 mm. Larger ones will restrict the volume of soil in which the young roots can grow and can contact soil surfaces. Silt-size aggregates are undesirable, because the resulting pores cannot be drained by gravity. Dispersed clay represents the most unfavorable structure.

Permeability To be desirable for plant growth, soil structure should be such that the infiltration capacity is large, the percolation capacity is medium, and air exchange is sufficient without being excessive.

Cohesiveness Cohesiveness of soil changes with moisture content. Therefore the first requirement to describe desirable cohesiveness is to specify that the soil should be in such a structural condition that it will be much of the time at the desired moisture level. This is the tension between field capacity and the wilting point. Soil should be in friable condition, but not too loose. Very loose soil suffers from excessive aeration and does not provide enough contact between roots and soil and does not give the plants sufficient support. Massive, compact soil restricts aeration and root spread.

It is desirable that the individual peds are highly cohesive, as this will protect them from destruction by rainfall impact or submergence.

MEASURING SOIL STRUCTURE

The four phenomena of soil structure—porosity, aggregation, cohesiveness, and permeability—can be studied by direct observation, or they can be quantitatively measured.

Table 5-3 Desirable pore-space distribution

Pore classification	*Pore sizes, mm in diam.*	*Corresponding soil-moisture tension, pF**	*Function of pores*	*Best pore distribution*					
				Semiarid climate	*Upland soil in humid climate*	*High groundwater in humid climate*	*Perhumid climate or irrigation in humid climate*	*Irrigation in dry climate*	*Golf green*
Large pores	>0.06	<1.7	Aeration and infiltration	Medium	Medium	High	High	High	High
Medium pores	0.06–0.01	1.7–2.5	Conduction of water	Medium	Medium	High	Medium	Medium	High
Small pores	0.01–0.0002	2.5–4.2	Storage of available water	High	Medium	Low	Low	Medium	Low
Hygroscopic surfaces		4.2–7.0	Storage of hygroscopic water	High	Medium	Low	Low	Medium	Low

* The pF units given correspond to the pore sizes only in cases of soils that are essentially free of soluble salts.

Qualitative field tests A rather valuable tilth diagnosis (determination of soil structure and its stability) can be performed with simple tools. The soil profile can be exposed with a spade or pick, and aggregate size, shape, and distribution can be observed. Placing some of the aggregates in water will show their stability, and pouring a bucket of water on top of the ground or on top of an exposed lower horizon will give a picture of the porosity and permeability of the soil. Another rapid test of soil structure can be performed by use of a soil probe (Hoffer tube). After preparing a straight cut of the profile in the probe with a knife, the entrance of a chalk suspension into the soil is used as an indication of the porosity. An eight- to ten-power magnifying lens serves to detect the shapes of small aggregates and the presence of clay skins. A soil penetrometer registers the pressure required to push a metal rod or disk into the ground and thus measures an aspect of cohesiveness.

Quantitative structure analyses The pore-space distribution of a soil can readily be determined by saturating a sample with water and removing it by use of successively increased tension. In this way pores from the largest sizes down to those of a diameter of 0.0002 mm can be determined. The volume weight of a soil whose true density is known permits the calculating of the total pore space.

The distribution of aggregates can be found by dry sieving; their stability and the size distribution of stable aggregates by wet sieving or by determining the settling rate of the aggregates in water. Aggregate stability can be measured also by the effect of water impact on individual aggregates or on the entire soil surface. The shape of the aggregates can be determined approximately by measurement of the vertical and horizontal axes.

To measure the various phases of permeability, a number of methods exist to determine the infiltration capacity, percolation capacity, the resistance to mechanical penetration and to air flow, and the consistence of the soil. The degree of shrinking of a soil paste is determined by placing it into a flat stainless steel cylinder and allowing it to dry out slowly. The decrease in diameter can readily be measured with a caliper.

A very valuable approach to soil-structure research is *soil micromorphology*, the study of soils with the aid of various types of microscopes (A. Jongerius, ed., 1964). In this way it is possible to determine sizes, arrangement, and percentage distribution of pores and of solids and to obtain an insight into the types of cementing materials. Observation of the shape and distribution of roots can serve as an indirect study of soil structure. In addition to the optical microscopes and the transmission electron microscopes, more recently the scanning electron microscope has been used for soil-structure studies. It provides reflected micrographs of

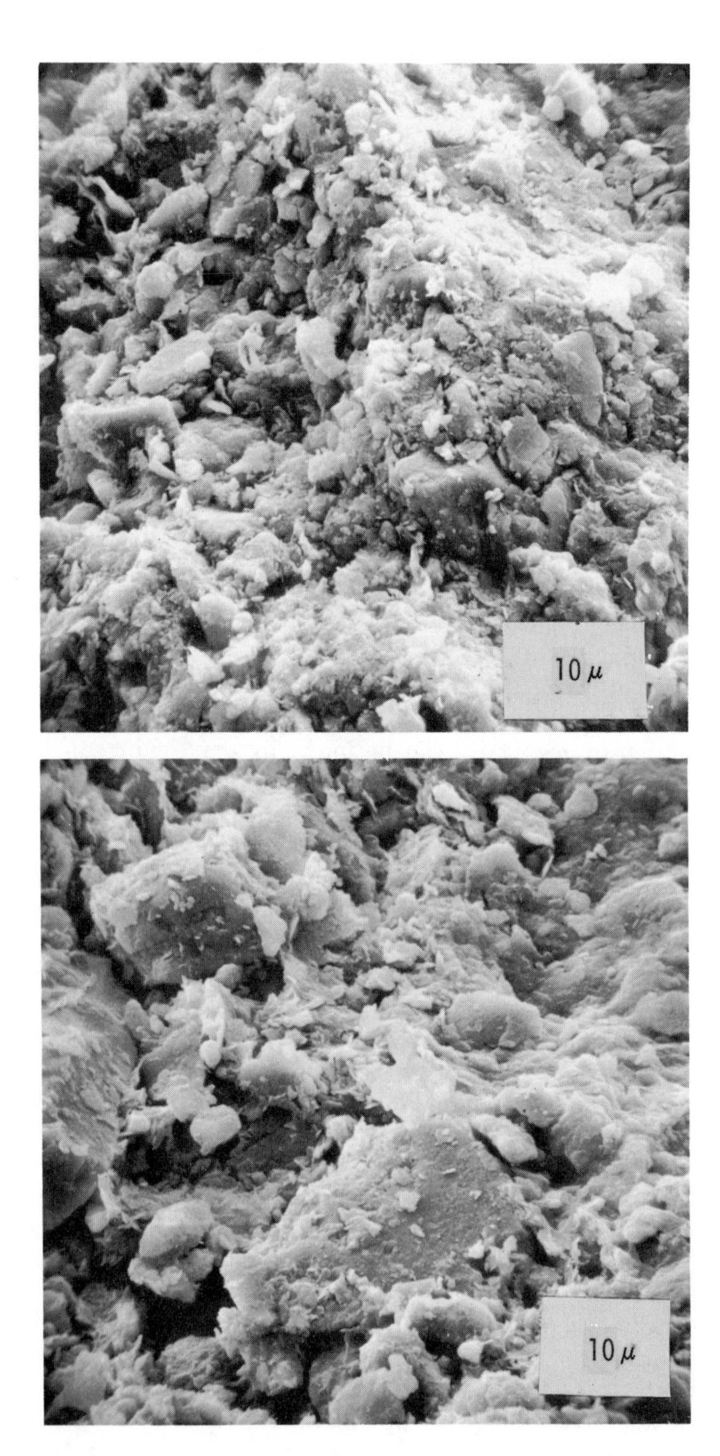

Fig. 5-7 Scanning electron micrographs of two silt loam soils: (*a*) Soil of friable, mellow structure. (*b*) Soil of cloddy, dense structure. (*Photographs by Engis Equipment Company, Morton Grove, Illinois.*)

considerable clarity up to 50,000 magnification. The great depth of focus of the pictures together with the possibility of stereoscopic view permits observation of soil particles down to colloidal size. Scanning electron micrographs of two soils of different structure are shown in Fig. 5-7*a* and *b*.

MANAGING SOIL STRUCTURE

Aims *Crop growth:* The aim of soil-structure management is to get porosity, aggregation, and permeability to air, water, and roots into as nearly optimal condition as possible, and to do this to a sufficient depth that the majority of the crop roots find such favorable conditions that the maximum yields possible under the given conditions of climate and plant nutrition result.

Soil conservation: The second aim of soil-structure management is to decrease its detachability and transportability by water or wind and to increase its infiltration and percolation capacities so that runoff and erosion are kept at a minimum.

Principal approaches In order to reach these aims several important approaches have to be followed: regulation of moisture, of aeration, and of consistence of the soil; deepening of the productive soil; erosion control; and destruction of horizontal structure units.

Methods Since most soils are flocculated—except some of the sodium soils—it is our task to separate masses of soil, not to aggregate them. Once aggregates of desirable size have been obtained they have to be protected from further breakdown which would eventually result in a massive soil.

The actual methods that follow these avenues to bring about desirable soil structure are manifold and varied. Most of them serve several purposes at once. The methods will only be enumerated here: proper land use, increasing plant growth, addition of organic matter, fertilization, tillage, subsoil improvement, mulching, drainage, irrigation, specific soil conservation methods, protection from compaction when wet, from rainfall impact, and from rapid frost, and possibly the use of soil conditioners. More detail on the management of soil structure will be found in Chap. 10.

REFERENCES

Atterberg, A.: Die Plastizität der Tone, *Intern. Mitt. Bodenkunde,* vol. 1, pp. 10–43, 1911.

———: Die Konsistenz und die Bindigkeit der Böden, *Intern. Mitt. Bodenkunde,* vol. 2, pp. 149–189, 1912.

Davidson, D. T.: Penetrometer Measurements, "Measurements of Soil Analysis," pt. 1, pp. 472–484, Am. Soc. of Agron., Inc., 1965.

Felt, E. J.: Compactability, "Methods of Soil Analysis," pt. 1, pp. 400–412, Am. Soc. of Agron., Inc., 1965.

Jongerius, A. (ed.): "Soil Micromorphology," Elsevier Publishing Company, Amsterdam, 1964.

Richards, L. A.: Modulus of Rupture as an Index of Surface Crusting of Soil, *Soil Sci. Soc. Am. Proc.*, vol. 17, pp. 321–323, 1953.

Sekera, F., and A. Brunner: Beiträge zur Methodik der Gareforschung, *Bodenk. Pflanzenern.*, vol. 29, pp. 169–212, 1942.

Sowers, G. F.: Consistency, "Methods of Soil Analysis," pt. 1, pp. 391–399, Am. Soc. of Agron., Inc., 1965.

Templin, E. H. (Chairman): Report of the Committee on Terminology Relating to Soil Consistence, of the Soil Science Society of America, *Soil Sci. Soc. Am. Proc.*, vol. 12, pp. 524–525, 1947.

6 ORGANIC MATTER

Organic matter is an integral part of every soil and one that affects its physical and chemical conditions to a much greater extent than its proportional share would indicate. Consequently an understanding of soil physics cannot be complete without a close consideration of the soil organic matter.

DEFINITION AND CLASSIFICATION OF SOIL ORGANIC MATTER

All organic substances in the soil, living or dead, fresh or decomposed, simple or complex compounds, are part of soil organic matter. This includes plant roots, residues of plants and animals in all stages of decomposition, humus, microbes, and any organic compounds. Animals that live in the soil are excluded from this definition and it would probably be best to exclude also living roots. On the other hand, living bacteria, fungi, and other microbes are included for the simple reason that it is essentially impossible to separate these from the rest of organic matter in the soil.

This consideration shows that a clear-cut definition of soil organic matter is difficult, if not impossible. For practical purposes soil organic matter can best be classified into residues and humus. The residues include dead parts of plants and animals and animal excreta in all stages of decomposition. The humus is the dark-colored soil organic matter that has fairly definite chemical and physical properties and that is not subject to as rapid a rate of decomposition as the residues.

ORIGIN AND FORMATION OF HUMUS

While there are different types of humus, all of them originate through humification of plant and animal residues in the soil. The contribution to humus formation from plants is much greater than from animals, although much humus originates as excreta of soil animals that feed mostly on plant material. By humification is meant a process of partial decomposition of organic substances and the synthesis of some compounds that are specific for humus. During humification some of the readily attacked constituents are oxidized and lost as water, carbon dioxide, and other gases. Some of the minerals are leached out. After a year of humification only a fraction of the original dry weight of the organic substance is left. Its composition is materially altered. Generally, the percentage of nitrogen and minerals is much higher than in the original material, the carbon content is slightly higher, but both oxygen and hydrogen have incurred the greatest losses. Humification is a process of slow and incomplete oxidation. For this reason roots are probably the prime source of humus, because they decompose much more gradually than the remains of the aerial parts of the plants. Lying on top of the soil, the latter are kept moist and well aerated and are thus exposed to rapid decomposition.

There are different types of humification. The broadest subdivision is into terrestric, semiterrestric, and aquatic groups. These are differentiated both by the type and amount of plant material available and by the rate and type of decomposition, since these are all affected by the relative amounts of water and air in the soil. The type of humus in the majority of agricultural soils is of terrestric origin. This type will be emphasized in this chapter.

Since temperature, moisture, and oxygen supply determine the type and the rate of humification, climate and topography have an outstanding influence upon the kind of humus produced. Generally speaking a climate with a moist warm summer and a cold winter is conducive to large accumulation of humus. This is so especially if the day length in the summer is long, as it is in subarctic areas. This allows for much photosynthesis and relatively little respiration and decomposition. In a hot, dry climate little plant growth occurs and the rate of decomposition is rapid; consequently practically no humus is found in the soil. Soils that stay too cold for appreciable plant growth throughout the year are also generally low in humus. An example are the soils on high mountains in the tropics just below the snow line.

Although the plants growing in an area are the result of climate and soil, they themselves have a strong influence upon the resulting humus. In this connection the most important properties of plants are their root distribution, their base status, their lignin content, and the presence or absence of water-repellent substances and of growth-promoting and growth-impeding compounds.

COMPOSITION AND PROPERTIES OF HUMUS

A clear picture of the composition of humus does not exist in spite of extensive studies for over 100 years. Therefore, only a generalized description will be given here. The most important ingredient of humus is probably a complex of amino acids and ligninlike substances. Waksman (1938) called this the *ligno-protein complex*. However, so far, only small quantities of ligno-protein have been identified in humus. But it is generally accepted that the amino acid-ligninlike complex in humus owes its origin to proteins and lignins in the plant substances (Jenkinson and Tinsley, 1959). Other compounds in humus are carbohydrates, including cellulose and hemicellulose, and a small percentage of fats, waxes, and resins. In addition, plant-growth substances as well as inhibitors occur in the various forms of humus. The proximate analysis of a typical humus reveals the following composition:

45% ligninlike compounds
35% amino acids
11% carbohydrates
4% cellulose
7% hemicellulose
3% fats, waxes, resins
6% others

A typical elemental composition of humus occurring in mineral soil is as follows:

Element	*Percent by weight*
Carbon	52–60
Oxygen	32–38
Hydrogen	3– 4
Nitrogen	4– 5
Phosphorus	0.4–0.6
Sulfur	0.4–0.6

In this case the carbon-nitrogen ratio is approximately 10. This is typical for humus in the surface soil in a great many localities. The C/N ratio is usually higher in the subsoil than in the surface soil. The so-called raw humus, especially in cold climates, has a much wider C/N ratio. Particles of neutral humus are as small as the finest clay. They have a high cation exchange capacity (200 to 400 milliequivalents/100 g). Because of its ability to react with cations, humus can be considered as acting as a weakly dissociated acid. Humus also absorbs anions, but releases phosphates much more readily than inorganic soil colloids do. For this reason the level of soil reaction (pH) is not as important for plant growth in organic soils as it is in mineral soils. Calcium humate is practically insoluble and forms water-stable complexes with clay. Hydrogen humate is also only slightly soluble, but it is readily dispersed and can move in the interstices of the soil in this condition. Sodium humate and ammonium humate are water soluble to a considerable extent. Black alkali soils owe their color to the solubility of sodium humate. The density of humus is between 1.3 and 1.5 g/cc. The heat of wetting of humus is high: 20 to 40 cal/g. When wetted, humus swells greatly and can absorb from two to six times its own weight of water, but after humus is thoroughly dried, it is difficult to wet again due to the fineness of the pores and the presence of water-repellent oils, waxes, and resins. In other words, rehydration of humus is a difficult and slowly reversible reaction, in contrast to rehydration of clay.

CLASSIFICATION OF HUMUS

Depending upon the type of vegetation and the environmental conditions, various types of humus are formed. The term "raw humus" is used to designate a stage of organic residues in which decomposition has been impeded by one or several of the following factors: low mineral content, low temperatures, insufficient aeration, or the presence of phenolic or other compounds that inhibit bacterial growth. The plant fiber is still recognizable in raw humus. Its color is more brown than black. It is typically found as the surface horizon in soils of low fertility, especially in cold climates. There is usually a sharp boundary to the mineral soil below.

The first stage in the development of true humus is the so-called nutrient humus. Its composition is somewhat between that of the residues and of true humus. This means that nutrient humus contains a good deal of easily decomposable compounds, such as sugars, starches, and soluble nitrogenous material. For this reason nutrient humus serves as an important source of energy for soil microbes. By far the largest part of the total humus of a mineral soil is the true humus or neutral humus. It has also been called the maintenance humus (Dauerhumus), because it persists a much longer time in the ground than the nutrient humus. Its color is black or nearly so. The boundary between the layer of true humus in the upper horizons and the mineral soil below is very gradual.

Various other classifications of humus have been suggested. Foresters, for instance, classify organic matter into two groups, mull and mor. Mull is the humus-rich layer of mixed organic and mineral matter. A soil profile containing mull has a gradual boundary from the predominantly organic horizons to the underlying horizons, probably because of intensive animal activity. Mor on the other hand is essentially unmixed organic matter that rests with little mixing on the mineral soil horizons. Although there is no direct equation possible, it can be said that mull consists largely of neutral humus while mor is similar to raw humus. It occurs mostly in coniferous forests.

Humus can also be classified according to the horizons in which it occurs and consequently according to the condition in which it is. In this way it is separated into *L*, *F*, and *H* layers. These letters stand for litter, fermentation, and humified layers. According to the general classification used at the beginning of this chapter the *L* layer is not actually humus, but consists mostly of residues. The *F* layer, also called "Förna," contains a fairly large percentage of nutrient humus and some decomposing plant fragments, whereas true humus, unrecognizable as to origin, predominates in the *H* layer.

Science has so far found it impossible to classify the kinds and components of humus on truly basic principles. There are several reasons for this. The main one is that humus cannot be separated from the mineral fraction of the soil without drastic changes in its composition. As a matter of fact some of the clay-humus complexes are so intimately joined together that there has been found no way to separate them without completely altering the composition of one or the other component. There are no sharp boundaries between the different kinds of humus. Besides, humus is in a constant state of change. The classification represented in Table 6-1 follows the generally accepted scheme of separation according to solubility. It is given only to the extent that it may be of value to judge the effect of the various components on the physical conditions of soils. Much more detail can be found in the specialized literature on humus, a few examples of which are listed at the end of this chapter.

DECOMPOSITION OF ORGANIC MATTER

Plant and animal residues decompose in or above the soil under a variety of conditions. The rate of decomposition and the end products formed depend on temperature, moisture, air, chemicals, and microbes. The

Table 6-1 Classification of the components of neutral humus

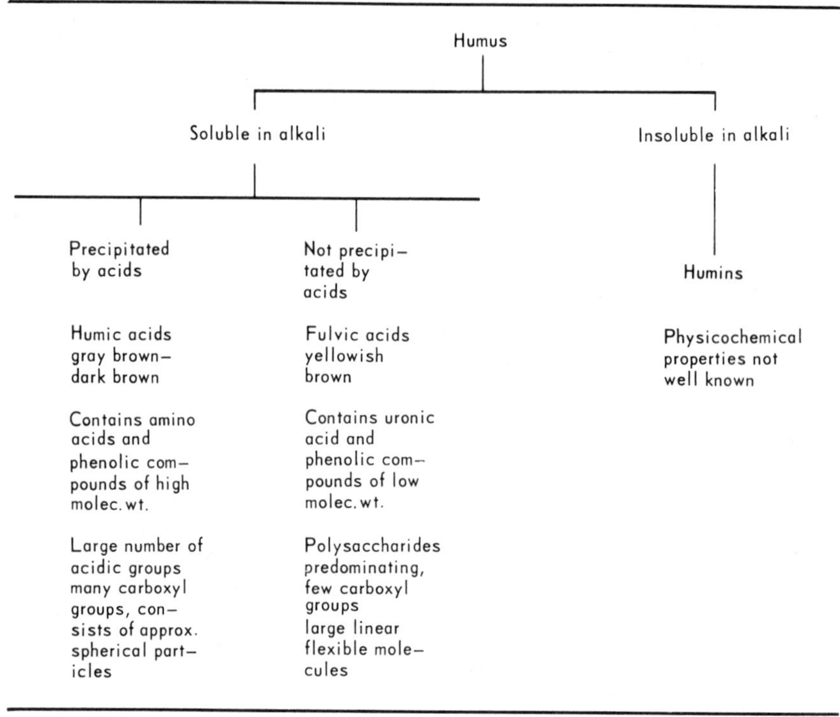

higher the temperature (up to 40°C), the faster the decomposition; that is the reason that tropical upland soils are generally low in humus. Moisture is needed for the biologic decomposition, but an excess of water causes a deficiency of air and therefore slows down decomposition.

Availability of the necessary chemicals for the nutrition of the microbes determines the rate of decomposition and influences the type of humus formed. Nitrogen is of prime importance in this respect.

Organic matter decomposes much faster in fertile than in infertile soil. The sequence of decomposition of the components of soil organic matter is generally:

1. Sugars, starches, water-soluble proteins,
2. Crude protein,
3. Hemicellulose,
4. Cellulose,
5. Oils, fats, lignin, waxes.

The rate of decomposition of organic matter decreases with time and with the attaining of a chemical composition similar to that of humus which can be regarded as one of the intermediate products of decomposition. The final decomposition products of organic matter are CO_2, H_2O, NO_3, SO_4, CH_4, NH_4, H_2S, etc., depending on whether the decomposition is aerobic or anaerobic.

Microbes and their enzymes are largely responsible for these processes. In fact, the decomposition of organic matter in the soil is a digestion process not unlike the digestion of organic matter in the stomach of an animal. Considerable amounts of oxygen are required for the decomposition of organic matter. It is clear, therefore, that oxidation of organic matter is most rapid in the surface soil, and much slower in the subsoil, especially if it is tight and wet.

A special case of decomposition of soil organic matter is the subsidence of muck and peat after they are taken under cultivation. These organic soils have developed in a condition of high groundwater, which has inhibited decomposition. In order to use such areas for crop production they have to be drained. This results invariably in a substantial increase in the rate of decomposition and consequently loss of substance. The very removal of water from muck or peat also causes some shrinkage. The subsidence due to these two reasons is substantial. It has been measured at between 2 and 5 cm/year in Florida and about half that amount in the northern states. This difference is due to the cold winters where organic matter decomposition slows down essentially to a standstill.

OCCURRENCE OF HUMUS

The amounts and types of humus occurring in and on soils depend on climate, vegetation, soil type, and topography. Maybe it would be more specific to say that they depend on temperature, moisture, aeration, day length, and the nutrient elements present.

A combination of a cool, humid climate and a soil rich in nutrients will result in the highest level of true humus. If the soil is poor in nutrients and only supporting pine forest, raw humus will predominate. In the hot climates of the tropical rain forests, soils are acid and poor, the organic residues decompose quickly, and the humus layer remains thin in spite of considerable plant growth.

Most of the humus occurs near the surface and gradually fades away until in the parent material no humus is found. The rate of decrease is gradual in grassland soils and much more abrupt in timber soils, because in the forest the majority of the residues occur on top of the soil, while grass roots make the major contribution to humus in the prairie (Fig. 6-1).

An exception to the gradual decrease in humus content with depth exists in podzolic soils in which some of the humus is dispersed as hydro-

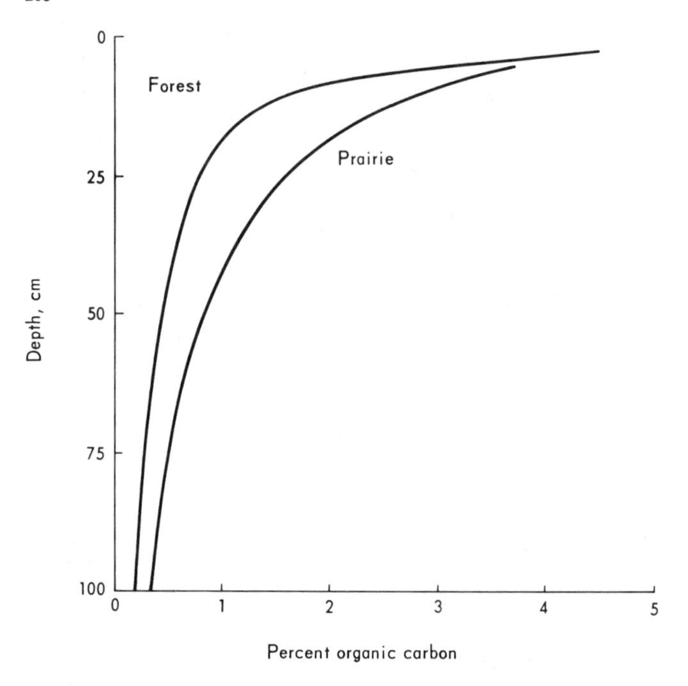

Fig. 6-1 Organic carbon profiles in Prairie (Brunizem) and Forest (Gray-brown podzolic) soils. (*After Wascher et al.*, 1960.)

gen humus and is leached downward. Depending on the pH profile, some of this humus will be precipitated at the boundary of the A and B horizons or, more frequently, at the bottom of the B, in the B_3 horizon, especially when an acid solum overlies a calcareous parent material. In such cases the B_3 is nearly saturated with bases and has a pH of 7.

In grassland the amount of "humus" is much larger than the amount of "residues." The richest chernozems contain as many as 1,300 metric tons of organic matter per hectare, considering the entire depth of soil. Forest soils have much less organic matter; about 50 to 180 tons/hectare. A considerable part of forest organic matter consists of residues, while the true humus frequently makes up less than half of the organic matter. The reason for this is that in a forest the majority of the organic residues are added to the soil from the top in the form of leaves, branches, and tree trunks. Roots contribute only a minor portion. On the other hand, in grassland there is a profusion of roots that add to humus formation. Since moisture and aeration have an influence on organic matter production and decomposition, it is easily understood that depressions usually are high in humus, whereas slopes, especially convex slopes, are low in humus. The amount of humus in cultivated soils depends on the humus originally present, the type of crop grown, and the fertilization and cultivation to which the soil has been subjected. Due to the increased aeration and the removal of parts of the crops, cultivated soils are lower in humus than they were before they were used by man. But exceptions exist.

FUNCTIONS OF ORGANIC MATTER IN THE SOIL

The functions of organic matter in the soil are manifold and in most cases beneficial for crop production and soil conservation. All these functions are interrelated.

Organic matter is a source of food and energy for microorganisms,
Organic matter assists in plant nutrition through its own decomposition and through the exchange capacity of the humus,
Organic matter provides the material needed for formation and stabilization of soil aggregates,
Organic matter improves the water-holding and water-conducting capacity of soils,
Organic matter in soils assists in the control of surface runoff and erosion.

The first two of these statements appear to be obvious. Plant materials containing sugars, fats, and protein furnish much energy and nutrition for microorganisms and release their stored mineral elements for use by plants and microbes. Assuming the third statement to be correct, the aggregation of soil particles provides for a soil structure containing both large and small pores and consequently improves water and air regimes. A better infiltration and percolation rate will reduce runoff and erosion and the stable aggregates are less easily detached from the ground and transported by the water.

It seems that the only one of the six statements of the functions of organic matter in soil that requires a detailed discussion is the third one: organic matter provides the material needed for formation and stabilization of soil aggregates.

While mineral soil particles, especially clay, can cling together to form aggregates or "peds," organic substances assist in this formation and give the aggregates the necessary stability to withstand the action of water.

In spite of a great number of studies, all the details of the interactions of organic matter and mineral soil particles are not yet known. It is only the true humus that participates in the formation of organic-mineral matter compounds. The "residues" in early stages of decomposition, while valuable in improving the structure of the soil, do not combine with the mineral fractions. Among the mineral fractions it is essentially only the clay, because of its large specific surface, that combines with the humus. Therefore, it might be logical to speak of clay-organic complexes. There is no doubt that humus also coats silt and sand fragments but the overall effect of this on the structure of soil is small.

In a soil with an organic matter content of less than 10 percent, by far the largest part of the true humus is tied to clay. There are two main types of mechanisms by which this is accomplished.

Generally speaking, the electrically charged humus components are absorbed chemically, while the uncharged ones are absorbed physically. The first group is primarily represented by the humic acid particles. These are spherical and carry a predominantly negative charge. They would be repelled by the likewise negatively charged clay were it not for polyvalent cations, especially Ca^{++}, and positively charged aluminum oxides that serve as bridges to link the two negatively charged colloids. Because of their round shape, humic acid particles have only few points of contact with clay and therefore are not held by physical forces. Their size is so large that they cannot enter the interlamellar spaces of expanding clays.

In the case of the uncharged humus components, particularly the polysaccharides that are part of the fulvic acid fraction, the adsorption force is physical. Long carbon-chain molecules are held closely to the surface of the clay by van der Waals' forces. Such molecules have to be long enough—at least a chain of six carbons or a molecular weight of 150—and flexible enough to have a large number of points that contact the surface of the clay. These molecules compete with water for the space close to the clay. For this reason drying out of soil induces a more secure adhesion of the polysaccharides to the clay. As a matter of fact, after drying out and warming up of the soil, the clay-organic contact is so intimate that a separation is not possible short of oxidation of the organic component.

Since the diphormic clays have very little electric charge, except at the broken edges, their organic complexing is due mostly to polysaccharides and van der Waals' forces. In the case of the triphormic clays ionic bonding predominates.

Much of the clay-humus complexing takes place in the guts of earthworms and of tiny animals in the soil. Here the contact of clay and humus particles is very intimate and bacterial reduction of fresh organic matter is intense. The result of the complexing of organic compounds with clay and other mineral components of the soil is the formation of aggregates and the protection of aggregates from dispersion by water. The effects of such soil structure on water and air in the soil and on the growth of plants and microorganisms have been discussed in Chap. 5.

Some of the organic coatings of soil mineral grains are hydrophobic and consequently hinder or even stop the wetting of the soil. This is particularly the case in sandy soils. Whether this is due to the greater drying in the sand or to the type of vegetation from which the organic matter originates or to some other reason is not known. It is known that the resinous compounds in pine needles are highly hydrophobic as is the residue of heather, two plant genera that frequently grow on sandy soils. Humus is truly the prime creator of compound soil structure. This reaction occurs mainly during formation or transformation of humus. Humus can absorb from two to six times its own weight of water, but its greatest value in increasing the water-holding capacity of the soil lies in the fact that humus improves soil structure. Humus improves every mineral soil; it gives cohesiveness to sands, mellowness to clays. It improves aggregation and porosity of the soil; it increases infiltration capacity and percolation capacity and thereby aeration and resistance to erosion. On the other hand, soils containing very much organic matter have little cohesiveness and low bulk density. This makes them erodible. Consequently mucks are greatly subject to wind and water erosion.

One of the most important functions of organic matter is that it supplies food and energy to the microorganisms. In this way the plant nutrients in the organic matter are released as well as acids which help to disintegrate soil minerals. This represents an important source of nutrients to the growing plants.

The beneficial functions of organic matter in the soil can be summarized as follows:

1. *Biologically:* Organic matter supplies energy, carbon, and minerals for microbes.
2. *Chemically:* Organic matter supplies carbon dioxide, nitrate, sulfate, and organic acids to help dissolve materials, and supplies other plant nutrients, both directly and indirectly.
3. *Physically:* Organic matter increases aggregation, protects the aggregates from destruction by water; makes the soil more tractable, and increases porosity and aeration; increases infiltration and percolation capacities; reduces runoff and erosion hazard.

ORGANIC MATTER AS A PARAMETER IN SOIL CLASSIFICATION

The official classification of soils in the United States makes a distinction between mineral and organic soils. The latter include the mucks and the peats. The boundary between organic soils and mineral soils is set at 30 percent organic matter for soils of high clay content (50 percent) and at 20 percent organic matter for soils essentially free of clay. For soils of intermediate clay content a proportional value is used as the limit of organic soils (Soil Survey Staff, 1960). Organic soils in which much of the structure of the original plant material is recognizable are considered to be *peat*, while those in which decomposition has proceeded further are called *mucks*. Normally peats are dark brown and mucks are black. This separation into organic and mineral soils does not provide for a differentiation of soils of organic matter contents below 30 to 20 percent. Yet it seems very important to convey the idea of the level of organic matter content along with the texture and geographic description. Such designations as "Yolo loam" or "Webster silty clay loam" tell the uninitiated nothing about the organic matter content of these two soils. Following the example of several European systems (Schlichting and Blume, 1966), it is proposed to classify the organic matter content of the "mineral" soils in the manner shown in Table 6-2.

Table 6-2 Classification of mineral soils according to organic matter content

Organic matter content, %	*Adjective describing humus content*
0–1	Very low humic
1–2	Low humic
2–4	Medium humic
4–8	High humic
8–20	Very high humic

In contrast to the adjective "organic," which is used for soils with more than 20 or 30 percent organic matter, the adjective "humic" is used to designate the organic matter content in "mineral" soils.

Speaking of a "low humic Yolo loam" or a "high humic Webster silty clay loam" gives a much clearer idea about the nature of these soils. Since the humus content of mineral soils has a great influence upon their water and air regime as well as on their behavior with respect to plant nutrition, such descriptive terms are definitely needed.

METHODS OF STUDYING SOIL ORGANIC MATTER

The study of soil organic matter is fraught with difficulties for various reasons. The humus is so intimately joined together with the mineral soil particles that it is impossible to separate the two without using strong solvents that modify the compounds present in the soil. On the other hand, it seems that humus is made up of a great array of compounds complexed together in different ways so that each soil may contain a different "humus." Nevertheless a number of analyses are available that can give a fairly clear picture of the organic matter in a soil (Kononova, 1961; and Black, ed., 1965). The determination of organic carbon and organic nitrogen reveals the total organic matter in the soil and the carbon/nitrogen ratio. This ratio helps one to recognize the state of decomposition of the organic matter and is important in planning soil management.

It is possible to separate organic matter that is bonded to mineral particles and that which is free (Greenland, 1965). This is done by flotation techniques, using liquids of densities around 1.8 to 2.0 g/cc. The free organic matter is floated off, while the mineral particles and the clay-organic complex sink to the bottom. Using liquids of densities of 2.2 to 2.5 g/cc it is possible to separate the clay-organic complex from the nonassociated mineral particles. This latter separation is not necessarily quantitative, because the relatively small amount of humus that can affix itself to silt or sand does not affect the density enough that a complete separation will be achieved. This portion of the humus, however, is minute.

A proximate analysis of humus into the main organic components can be achieved with a fair degree of accuracy. It is also possible to analyze the soil for the functional groups occurring in humus, for instance, carboxyl group, aliphatic hydroxyl group, phenolic hydroxyl group, and carbonyl group.

The classical approach to soil organic matter study has been to separate its components according to their abilities to dissolve and precipitate in solutions of bases and acids. This is the technique used to distinguish between humic acid, fulvic acid, and humins, which were mentioned in Table 6-1.

Instead of removing the organic matter from the mineral fraction of the soil, more recently a technique has been used to dissolve the mineral matter and to filter it off from the organic matter. This is done by replacing the exchangeable ions with hydrogen and by using hydrofluoric acid to dissolve the siliceous fraction. Although this treatment may also affect the organic components, it provides an additional approach to the isolation of humus.

Other methods of studying organic matter are to determine its color, to determine its state of physical decomposition by microscopic differentiation into optically amorphous material and into recognizable plant remains, and to determine its electric charge. A method of separating residues from humus is to treat the mixture with acetyl bromide. Residues dissolve in it, humus does not.

With the constantly improving analytical techniques, there is little doubt that before long many more of the properties of soil organic matter will be discovered.

MANAGEMENT OF SOIL ORGANIC MATTER

The overall purpose of managing soil organic matter is to increase crop yields and to conserve the soil. The type of management depends on the nature of soil and climate and the intended land use. The goal is to have a generous amount of nutrient humus and stable (neutral) humus in the soil. This requires production of large amounts of organic matter and distribution of it through as deep a profile as possible. This implies liming and fertilization and the growth of crops that leave ample residue material above and within the soil. The fibrous root systems of the grasses and the deeply penetrating roots of some of the legumes are particularly valuable in this connection.

Nitrogenous compounds and lignin are important in the production of humus. Sufficient amounts of nitrogen have to be present to combine with the carbonaceous components of the residues for maximum humus formation. As long as the carbon/nitrogen ratio of organic matter is larger than 30:1, nitrogen is taken from the supply of mineral nitrogen in the soil. Since the benefits of organic matter are particularly valuable in the surface layer of the soil, the main accumulation ought to be here. Incorporation in the upper 5 cm of the soil is preferable over leaving too large a proportion of the residues on top of it. It has been found that surface application of organic residues (mulching) results in a lower oxygen content in the soil than any other method of application (Epstein and Kohnke, 1957). On the other hand, some soils with a heavy clay layer in the subsoil can be greatly improved by deep application of organic residues. In this way permeability for water and air are improved and the plant roots have a larger volume in which to forage for nutrients and water.

One important consideration in the management of organic matter is that crop residues should be allowed to decompose at such a rate and during such a time that the main requirement for nitrogen and other nutrients of the microbes decomposing the residues does not coincide with the main requirement for nutrients by the growing crop.

Leisurely decomposition of residues results in the formation of the largest amounts of the most valuable humus. Rapid oxidation produces carbon dioxide, water, and other simple compounds that are of no value in humus production, instead of the large organic molecules that are the building blocks of humus. In the absence of free oxygen, on the other hand, reduced, poisonous compounds are formed.

If organic matter of wide carbon/nitrogen ratio (larger than 30:1) is used, one or more of the following practices should be adopted.

1. Incorporate crop residues early into the soil so that decomposition is partially completed before crop plants start growing.

2. Apply extra nitrogen (however, this does not prevent temporary oxygen deficiency).
3. Haul off the residues and use as bedding to incorporate manure before using the residues in the field.
4. Haul off and permit partial decomposition to take place before using the material in the field (compost or artificial manure).

In the management of crop residues, the needs of the crop plants and the importance of soil conservation must always be considered. This means that the actual treatment will be greatly affected by climate, soil, topography, and crop. In a dry climate and on well-aerated soils, a lowering of the oxygen diffusivity is of little consequence, whereas on less pervious soils or in a wet climate it could result in substantial yield decreases. Under extreme cases of organic matter accumulation, as they occur, for instance, in subarctic regions, only highly decomposed organic matter should be applied, since fresh residues will convert so much of the mineral soil nitrogen into organic compounds that the plants will starve. In general, however, there is little cause for concern that too much organic matter be used. It is only the problem of the right type and time of application. From the viewpoint of soil conservation it is always preferable to leave some of the crop residue on top of the ground, as it will protect the soil from the impact of the raindrops and thus reduce runoff and erosion. For the reasons stated above as well as because of weed control, it is frequently better to incorporate part or all of the crop residues into the soil. Since organic matter in the soil gradually decomposes whenever the temperature is above 5 to 10°C, it is necessary to replace it in order to continue to have the benefit of clay-humus complex formation. No recipe can be given as to the exact amount that needs to be added to the soil to maintain its structure.

Relatively small annual additions are sufficient for this purpose. Rates of over 5 metric tons/hectare-year are not required. As a matter of fact, they can be disadvantageous. Reduced conditions may result, the water balance of the soil is disturbed, and crop yields suffer. Two to four tons of organic residues (on dry-matter basis) per hectare-year is sufficient under most agricultural conditions.

In the majority of cases the organic matter available for application on a given field are the crop residues that have grown there. The problem is in which way this resource should be used to give the greatest benefit. Essentially there are four possibilities. The residues can be left on top of the ground as mulch. They can be mixed with the very surface soil—maybe to a depth of 5 to 7 cm. They can be incorporated into the plow layer—about 20 to 25 cm. And finally, they can be placed in vertical slits into surface soil and subsoil. This latter technique is called *vertical mulching*.

The first method is used where erosion is a major problem and where excess moisture in the soil is not likely to occur. Mixing the residues with the surface layer of the soil is particularly valuable where crusting is a hazard. This method not only provides a good seedbed, it also helps to increase the infiltration capacity. Incorporating the residues throughout the plow layer is the classical method and is practically unavoidable where the plow is used.

Unless there are specific reasons for another form of application, incorporation throughout the plow layer is the most efficient overall use of crop residues. Vertical mulching, when properly done, helps in getting rainwater into the soil and decreasing surface runoff rates and erosion. It is useful where subsoils are dense and poorly permeable for water. So far, however, vertical mulching is still in an experimental state.

In the management of organic soils for crop growth, excessive oxidation is an ever-present danger because of their great aeration porosity. Such soils have to be drained to provide the roots of the mesophytic crop plants with a favorable habitat. This drainage is the cause of loss of muck and peat soils and should always be done with caution. On such organic soils the water table should be just deep enough to allow the plants to thrive during the vegetation period. During the rest of the year the water level should be at the surface of the soil.

As a reaction to the greatly expanding use of commercial fertilizers, a group of people have expressed the opinion that "artificial" fertilizers are poisonous for plants and for the animals and men that eat these plants and that only organic fertilizers should be used. It is claimed that this "organic farming" results in healthier plants, less plant diseases and pests, less weed growth, and larger yields. There is no doubt that a large amount of well-prepared compost can be of great benefit to crop plants and to the conservation of the soil. It can also be assumed that compost or organic residues in general will provide the chemicals in a proportion similar to the one needed by the plant, while in the use of commercial fertilizer excessive availability of one or the other nutrient is a possibility. On the other hand, it must be recognized that organic residues are completely insufficient to fertilize all agricultural land to a level where large yields can be obtained year after year. The main value of the "organic farming" movement seems to consist in pointing to better plant-residue management methods and to the possibility of accumulation of toxic substances in the soil.

REFERENCES

Black, C. A. (ed.): Chemical and Microbiological Properties, "Methods of Soil Analysis," pt. 2, Am. Soc. of Agron., Inc., 1965.

Epstein, E., and H. Kohnke: Soil Aeration as Affected by Organic Matter Application, *Soil Sci. Soc. Am. Proc.*, vol. 21, pp. 585–588, 1957.

Greenland, D. J.: Interaction between Clays and Organic Compounds in Soils, *Soils and Fertilizers*, vol. 28, pp. 415–425, 521–532, 1965.

Jenkinson, D. S., and J. Tinsley: Studies of the Organic Material Extracted from Soil and Compost, *J. Soil Sci.*, vol. 10, pp. 245–263, 1959.

Kononova, M. M.: "Soil Organic Matter, Its Nature, Its Role in Soil Formation and Soil Fertility," English translation, T. Z. Nowakowski and G. A. Greenwood, Pergamon Press, New York, 1961.

Scheffer, F., and B. Ulrich: "Lehrbuch der Agrikulturchemie und Bodenkunde, III, Humus und Humusduengung," Ferdinand Enke Verlag, Stuttgart, 1960.

Schlichting, E., and H. P. Blume: "Bodenkundliches Praktikum," 209 pages, Paul Parey, Hamburg and Berlin, 1966.

Soil Survey Staff: "Soil Classification, a Comprehensive System," 7th Approximation, Soil Conservation Service, U.S. Department of Agriculture, 1960.

Waksman, S. A.: "Humus, Origin, Chemical Composition and Importance in Nature," 2d ed., The Williams & Wilkins Company, Baltimore, 1938.

Wascher, H. L., J. D. Alexander, B. W. Ray, A. H. Beavers, and R. T. Odell: Characteristics of Soils Associated with Glacial Tills in Northeastern Illinois, *Univ. of Illinois Agr. Expt. Sta. Bull.* 655, 1960.

7 SOIL AIR

FUNCTIONS OF THE INDIVIDUAL COMPONENTS OF SOIL AIR

The various components of the soil atmosphere are just as necessary to the productivity of the soil as the solid and liquid phases. Oxygen is required for the respiration of plant roots, microbes, and the soil fauna. Carbon dioxide helps to dissolve nutrients and to make them available to the plants. Nitrogen gas serves for the production of combined nitrogen by symbiotic and nonsymbiotic bacteria. Water vapor prevents the desiccation of roots and microbes and aids in the transfer of water within the soil.

Of particular importance is the presence of adequate amounts of oxygen, as it is constantly exhausted by roots and microbes. Without sufficient oxygen in the soil the normal functions of most crop plants and of the aerobic microbes come to a standstill. Anaerobic bacteria use oxygen in organic and inorganic compounds, reducing them to sulfides, nitrites, ferrous compounds, and other reduced compounds that are toxic to the plants. An excess of oxygen in the soil is also undesirable because the organic matter would be oxidized too rapidly. Semi-aerobic decomposition is best for the production of the largest amount of true humus and for the steady supply of organic compounds that serve to stabilize soil aggregates.

PLANT REQUIREMENTS FOR SOIL AIR

For the respiration of plant roots a constant supply of oxygen is needed. The more plants per acre and the more vigorously they grow, the more oxygen is required. Normal growth of most crop plants is possible only if the oxygen concentration exceeds 10 percent.

As soon as the oxygen supply in the soil is limited, the rate of growth of most crop plants slows down and stops entirely when the oxygen concentration sinks below 2 percent. An ample supply of oxygen in the soil allows the plants to use the other nutrients and water more efficiently. For instance, it allows the plants to have a lower water requirement. A low oxygen concentration of soil air interferes with water uptake (physiologic drought). The effect of aeration on nutrient uptake is different for the different elements. This effect is greatest for potash uptake and decreases in the sequence calcium, magnesium, nitrogen, phosphorus (Lawton, 1945). It seems that aeration is the most important factor determining rooting depth. Generally speaking the thickness of roots determines the oxygen-pressure requirement, which means that thick roots require higher oxygen concentration in the soil air than thin ones.

The aeration requirements for the growth of the various plants vary considerably. Tomatoes, potatoes, sugar beets, peas, and barley have very high oxygen needs. Corn, wheat, oats, and soybeans belong to the next group. Many grasses seem to have lower requirements than corn and wheat. Sorghum can survive flooding for several days without ill effects. Sudan grass and reed canary grass get along on very restricted aeration. Willows, rice, cattails, and some sedges can pump air down into the roots. The ability of rice to thrive under flooding is attributed to the presence of interconnecting air chambers in the cortex, which are able to supply oxygen to the roots so long as the top remains exposed to the atmosphere.

It has been found that plants that are actively thriving suffer greatly and may die when their oxygen supply is restricted. On the other hand, plants can survive low oxygen levels if they are gradually preconditioned.

It is difficult to determine the optimum and minimum requirements of soil oxygen by the various plants. It seems that most plants grow well in a soil atmosphere of considerably less than the 21 percent O_2 of the free atmosphere. Crop plant growth is restricted when the oxygen content of the soil air sinks below 10 percent. The important feature seems to be the continued supply of oxygen, more than the actual partial pressure. The oxygen diffusion rate should be at least 30×10^{-8} g/cm^2-min for satisfactory growth (Bertrand and Kohnke, 1957; Stolzy and Letey, 1964). It is generally agreed that it is not the excess of carbon dioxide that hurts the plants—as long as it is within reasonable limits (less than 20 percent CO_2)—but the lack of oxygen. Excessively heavy oxygenation tends to retard the growth of corn and cotton.

COMPOSITION OF SOIL AIR

Compared with atmospheric air, the content of carbon dioxide and water vapor is higher in soil air, that of oxygen is lower, and the nitrogen content is about the same in both. The sum of the oxygen and the carbon dioxide contents in the soil is nearly the same as in atmospheric air. It is, however, not constant. In the surface layer of a "well-aerated" soil, the oxygen content is between 18 and 21 percent, while at a greater depth and especially in soils that are wet for a long period it may be very much lower. Under reducing conditions soil air may contain methane, hydrogen sulfide, and ammonia. The CO_2 content of soil air is usually between 0.1 and 5 percent and can reach nearly 20 percent.

The composition of soil air is subject to rapid fluctuations since it is the resultant of two dynamic properties: the change of oxygen to carbon dioxide by roots and microbes and the renewal of soil air by atmospheric air.

The effects of various factors on the composition of soil air can be summarized by the following statement:

Other things being equal,
the CO_2 content is higher in summer than in winter, because of greater root and microbe activity;
the CO_2 content is higher in manured, limed, fertilized, and vegetated soil than in untreated, unvegetated soil, for the same reason;
the CO_2 content is higher in wet than in dry soil, because of restricted diffusion;
the CO_2 content is higher in fine-textured than in coarse-textured soils, because of restricted diffusion rates, usually resulting from higher moisture contents;
the CO_2 content is higher in poorly aggregated or puddled soils than in soil with crumb structure, again because of the difference in diffusion rates;
the CO_2 content is higher in the subsoil than in the topsoil because the surface soil diffuses directly against the atmosphere, while the subsoil diffuses against the surface soil.

In the above statements the words "the CO_2 content is higher" can be replaced by the words "the O_2 content is lower."

AIR CAPACITY OF THE SOIL

The actual air capacity of a soil is that part of the pore space which is filled with air and not with water. This value changes constantly with the moisture content. It also changes with the structure. Generally soils with low groundwater table and coarse texture have high air capacities. In order to know what amount of air capacity to expect in a soil, it is well to determine its *aeration porosity*. This is defined as the pore space filled with air when the soil is under a tension of a 50-cm column of water. This value is also called *aeration capacity* or *noncapillary porosity*. It corresponds to a pore size of 0.06 mm in diameter or larger. No agreement has been reached among the various workers concerning the exact height of the water column providing the tension. Values from 40 to 100 cm are used. The meaning of "aeration porosity" is to set up a quantitative standard by which the potential air capacities of soils can be compared. A soil of a given aeration porosity will reach an air capacity equal to the aeration porosity a few hours after an excess of water has had a chance to drain off. If the groundwater level (it may be temporary) is less than 50 cm below the soil in question, the air capacity will be smaller than the aeration porosity, because the tension is too small to remove the excess water from all pores larger than 0.06 mm in diameter.

Experience has shown that a soil generally is well aerated for crop plants if it has an aeration porosity of 10 percent or more of the total volume, unless of course there is a high groundwater table. It seems that permanent pastures have a somewhat lower aeration-porosity requirement.

The aeration porosity of medium- and fine-textured soils depends largely on the state of aggregation. Soils with crumb structure have greater aeration porosities than those with angular aggregates or with single-grain structure. Sandy soils have higher aeration porosities than loamy or clayey soils. Latosols, even those with high clay contents, have high aeration porosities, because of their pronounced aggregation. Mucks have very high aeration porosities, usually above 20 percent.

Soil pores that do not assist in aeration in spite of sufficient size because they are tightly surrounded by smaller pores are called *blocked pores*. They may remain filled with air even when the soil is submerged because the pores around them are so small that they prevent the escape of the air.

RENEWAL OF SOIL AIR

The renewal of soil air goes on both as diffusion of the various gases and as mass flow.

DIFFUSION

Numerous studies have shown that diffusion is by far the most important means of renewal of soil air. Diffusion is the random movement of molecules of a gas or of a liquid. Net movement of gases by diffusion occurs when the partial pressures of individual gases in two neighboring systems are different, but the total pressure is the same in both. Diffusion is proportional to the square of the absolute temperature.

The rate of gaseous diffusion in soils is a linear function of the effective pore space. This relationship can be expressed by the equation

$$D = D_o k S$$

where D = coefficient of gaseous diffusion through soil of effective pore space S
D_o = coefficient of diffusion through still air
k = soil diffusion constant

By "effective pore space" is meant that part of the total pore space that permits free passage of gases, i.e., the pore space that is not filled with water and is not blocked off from other pores by small water-filled pores.

The soil diffusion constant k expresses the reduction of the rate of diffusion in the soil pores compared to that in free air. In the soil the gas molecules have to travel tortuous paths that are longer than the thickness of the soil layer. It is natural that these detours slow down the progress of the gases. In addition, the pore diameters change from spot to spot and this also impedes diffusion. The actual value of k depends on the structure and the moisture content of the soil. It is usually between 0.6 and 0.8. The reason that diffusion through soil is not reduced more is that the mean free path of gas molecules (0.0005 to 0.0001 mm) is much shorter than the diameter of even the smallest pores that can be free of water in a soil at field capacity (0.001 mm).

Since the rate of gaseous diffusion in soil depends both on structure and moisture content, it is not possible to designate a definite matric moisture tension as the limit of adequate aeration. If we specify a structure that is not compacted and is free of crust or pan, it can be said that a soil is well aerated when its matric moisture tension is more than 1 atmosphere.

Diffusion operates continuously, as long as there is effective pore space in the soil, whereas the agencies that cause mass flow of air function only intermittently.

MASS FLOW

Viscous or mass flow can also help to renew the soil atmosphere by changes in barometric pressure and in temperature, by change in soil-moisture content, and by wind. The effects of these phenomena of air renewal, however, are intermittent, and it can readily be determined that the total amount is inadequate to supply plant roots with the necessary oxygen and to remove the accumulated carbon dioxide from the soil, except for its very top layer.

A change in barometric pressure of 25 mm of mercury during one day is exceptionally large. Yet this is only one-thirtieth of the total pressure. According to the gas law, this would account for an expansion (or contraction) of one-thirtieth of the volume of the gas. Assuming an effective soil-air depth of 1 m, only the upper 3.3 cm of soil air would be exchanged against atmospheric air. Actually this figure is much smaller, for the upper soil normally has a much greater total and effective porosity than the subsoil and the daily barometric changes are normally smaller than 25 mm of mercury.

Daily temperature fluctuations seldom reach the depth of 1 m. Below 50 cm such fluctuations are insignificant. In the upper 10 cm there are daily temperature changes up to 20°C, but as an average of the upper 50 cm a fluctuation of 5°C is greater than usual. Yet 5°C is only five three-hundredths or one-sixtieth in terms of the absolute temperature. Again, a daily gas exchange of one-sixtieth of the top layer of the soil air is completely inadequate to supply the plant roots with oxygen.

As water drains out of the soil or evaporates from the soil surface, fresh air is pulled into the soil; the volume per day involved in this exchange is small indeed.

No specific information is known to determine the effect of wind on soil aeration. In wind-erosion studies, however, it has been determined that the wind velocity at the very soil surface is seldom large and in a field covered with dense vegetation it is zero.

As mass flow cannot possibly be of much importance in the renewal of soil air, the laws of mass flow of gases are of little significance in studying soil air. Since mass flow is viscous, the size of the pores through which the air passes is essential. As a matter of fact, the rate of mass flow of gases is proportional to a higher power of porosity.

RAINWATER

Another source of oxygen is that one dissolved in rainwater. Assuming that rain is saturated with oxygen, which is essentially true, 1 cm of rain over an area of 1 hectare or 100 m^3 (100,000 liters) contains 4,339 g of oxygen at 20°C. This is the equivalent of about 3,000 liters of pure oxygen at atmospheric pressure. Consequently a rain usually has a much more invigorating effect on a crop than does an irrigation.

SOIL AERATION AND OTHER SOIL PROPERTIES

Probably the outstanding effect of soil-aeration conditions—if we exclude the effect on plant growth—is upon microbe activity and indirectly upon other soil properties. A vigorous growth of aerobic bacteria requires the presence of sufficient oxygen, water, energy material, and mineral nutrients. Under such conditions large quantities of organic compounds are formed that serve to hold sand, silt, and clay together and to coat such aggregates. The result is a water-stable structure.

Lack of free oxygen results in presence of insufficient amounts of the type of organic compounds that are needed to stabilize soil. For this reason, well-drained, well-aerated soils have usually a more favorable structure than poorly drained, poorly aerated ones.

The organic matter content as such is also influenced by aeration. Excessive aeration, as it occurs in many sandy soils and in dry climates, causes organic matter to be oxidized quickly. Generally speaking, the slower the aeration of a soil, the higher its organic matter content will be. For this reason areas that are continuously under shallow stagnant water and are therefore poorly aerated frequently form muck and peat soils. Lack of aeration results in the formation of methane, hydrogen sulfide, nitrite, ammonia, free nitrogen, and ferrous compounds. Some of these and other compounds developed under these conditions are poisonous to plants.

If a soil is poorly aerated through many years, either seasonally or continuously, iron will be leached out in the ferrous form, in which it is much more soluble than in the ferric form.

MEASURING SOIL AERATION

The task of determining whether a soil is adequately aerated for best plant growth or not is a difficult one. The safest way of doing it is to set up a vegetation experiment in which aeration is the only variable—as far as this is possible. The difficulty is to maintain constant aeration conditions throughout long enough a period so that the effect on the plants can be measured. In such an experiment it has to be decided whether the plant response is to oxygen pressure, carbon dioxide pressure, moisture tension, or the presence of a poisonous substance.

In order to study the aeration conditions of a specific soil, it is preferable to determine individual soil properties that represent aeration or are related to aeration. Such determinations are described below.

GASEOUS DIFFUSION RATES IN SOIL AT VARIOUS SOIL-MOISTURE TENSIONS

There are several methods of determining diffusion rates in soil. The one most frequently used is probably the platinum electrode technique (Lemon and Erickson, 1952). An electric potential is set up between a platinum electrode in the soil and a reference electrode. Oxygen reduced at the platinum electrode depends on the amount of oxygen diffusing to that electrode through the moisture layer surrounding the electrode and at the same time determines the amount of electric current. Thus it is possible to estimate the rate of oxygen diffusion as it would reach a plant root (Letey and Stolzy, 1964).

In another, more direct method a small chamber with a closable opening at the lower side is buried in the soil. Two tubes lead from it to the surface. With the bottom valve closed, the chamber is flushed with nitrogen, the tubes are closed, and the valve opened again. After a given time the contents of the chamber are analyzed for oxygen. In this way the rate of diffusion from the surface can be estimated (Raney, 1949; Robinson, 1957).

In order to determine the ability of a soil sample to permit gaseous diffusion, it can be fitted into a special box containing carbon disulfide in its lower part, while the soil rests on a fine screen above it. The soil is sealed into this chamber in such a way that the carbon disulfide can escape only in vapor form through the soil. Assuming that the soil was air dry at the beginning, any loss in weight of the system is due entirely to the evaporation of carbon disulfide that has diffused through the soil (Domby and Kohnke, 1956). Determination of diffusion through a moist soil by this technique is much more complicated, since a change in moisture content has to be avoided.

AMOUNT OF CARBON DIOXIDE GIVEN OFF AT THE SOIL SURFACE

Various methods exist to determine the carbon dioxide given off from a definite area at the soil surface. This gives a picture of the microbial and root activity in the soil.

COMPOSITION OF SOIL AIR

With modern equipment the oxygen content of soil air can be determined very rapidly (van Bavel, 1965). However, only the air from the larger voids is sampled and not the air that is in contact with the majority of the root hairs. Nevertheless such a determination gives a good idea of the aeration condition in the soil.

RESISTANCE TO MASS FLOW OF AIR UNDER PRESSURE

A small cylindrical chamber is pressed partially into the soil surface and air pressure is applied. The resistance of the soil to air flow can be determined with manometers. Janert (1937) and Wilde and Voigt (1955) have designed practical methods for this purpose. Since mass flow is of little consequence to soil-air renewal, this type of determination seems to have only small value in the study of soil aeration.

AERATION POROSITY

Determination of aeration porosity—also called aeration capacity or noncapillary porosity—gives an idea how much of the soil volume can be expected to be filled with air a few hours after a heavy rainfall. Aeration porosity is the total volume of pores that have a diameter larger than 0.06 mm. It can be determined by applying tension equivalent to a water column of 50 cm to a saturated soil (tension table). Generally if the aeration porosity amounts to 10 percent or more, aeration is satisfactory unless there is a very shallow groundwater table (less than 50 cm).

AIR CAPACITY

While aeration porosity is a property that is determined without regard to the moisture content at the time of determination, air capacity or effective air capacity is that amount of pore space that at the moment is filled with air. It is therefore a very dynamic property, changing not only with the structure, as aeration porosity does, but also with the moisture content. A cylindrical sample of soil is taken in its natural structure and its gas content determined immediately with a portable air pycnometer in the field (Nietsch, 1936; and Russell, 1949). A special analysis is required to determine the volume of blocked pores that do not contribute to aeration.

OXIDATION-REDUCTION POTENTIAL

The value of the oxidation-reduction potential in determining the aeration conditions of soils has been much discussed. Since the oxidation-reduction potential in most soils is poorly buffered, it is difficult to determine it reliably, especially if a sample is to be taken in the field and brought to a laboratory. Another problem is whether the oxidation-reduction potential as such actually represents the oxidation-reduction situation in a soil, because some of the organic substances are not truly reversible and therefore do not affect the potential in the same way as, for instance, the ferric-ferrous system. There is no doubt, however, that the determination of oxidation-reduction potentials can be very helpful in some types of soil research. This can be determined electrometrically or by oxidation-reduction color indicators.

AMOUNT OF REDUCED MATERIAL IN THE SOIL

When soils have suffered from poor aeration for some time, a variety of reduced substances accumulate. This can be determined quantitatively by titration. It is expressed as milliequivalent reduced material/100 g of soil.

SOIL-AIR MANAGEMENT

THE AIMS OF SOIL-AIR MANAGEMENT

As a generous supply of oxygen is required for best plant growth and highest yield of the crop plants, it is necessary to provide adequate aeration of the soil. Adequate aeration can be defined as a condition in which the oxygen diffusion rate is at least 30×10^{-8} g/cm^2-min and the oxygen concentration of the soil air at least 10 percent down to the depth of the genetic root limit of the crop in question. In an agriculture in which plant nutrients are applied generously in the form of fertilizers and water can be applied by irrigation, crop yields are frequently limited by the oxygen supply in the soil. On the other hand, excessive aeration has the disadvantage of tending to oxidize the organic matter too rapidly and drying out the soil. Under practical field conditions such excessive aeration occurs only on coarse-textured and highly organic soils and in areas of continued high temperatures during periods of drought. It may also result from overly frequent cultivation.

On medium-textured and heavy soils in humid climates, the problem is practically always to increase aeration. A good rule of thumb is to maintain 10 percent aeration porosity in the soil.

METHODS OF SOIL-AIR MANAGEMENT

Since diffusion is the main agent of air renewal, the methods of soil-air management have to be such that they affect potential diffusion rates. These are changes in structure, moisture content, and temperature. A more open structure, a lower moisture content, and a higher temperature will increase diffusion rates.

Since it has been established that the total volume of air-filled pore space determines the rate of diffusion and that the size of the pores has very little effect on it, it is necessary to discuss structure and moisture together. A soil crust or a dense plow layer permit ample diffusion of oxygen and carbon dioxide as long as they are dry. The important point is that the fine pores of such soils retain water a long time after a wetting and therefore inhibit diffusion. For this reason it is important to avoid crusts and compacted soil. Tillage, incorporation of organic matter, and mulching are the standard methods to produce a soil of good tilth. All the means that serve to create a well-aggregated stable soil structure also serve aeration. With these methods it is relatively simple to prepare favorable conditions in the plow layer. For the improvement of the deeper layers of the soil we have to resort to the use of deep-rooting crops, or to subsoiling, subsoil fertilization, and vertical mulching (the incorporation of organic residues into slits cut into the subsoil).

Where soil aeration is restricted by a high groundwater table (permanent or temporary), drainage is an obvious remedy. This will not only decrease the moisture content and increase the effective air capacity but also increase temperature and indirectly affect soil structure.

Although an increase in soil temperature increases diffusion rates, it also increases microbial activity and hence carbon dioxide production in the soil. The net result of temperature increase on the partial pressure of oxygen in a soil may therefore be positive or negative. In this connection it is interesting to consider the effect of mulch on aeration. Mulch protects the soil from the impact of the raindrops and therefore protects its tilth. This favors aeration. On the other hand, it keeps the soil more moist, thus restricting the effective air capacity. During spring and summer when much aeration is needed, it keeps the soil cool, thus decreasing diffusion rates. Whether the total result of mulching on diffusion will be positive or negative depends upon the individual case. However, observation indicates that it usually decreases diffusion of oxygen into the soil.

Other means of managing soil aeration include regulation of respiration of roots and microbes by fertilization and cultural practices, density of plant stand, amount and location and time of incorporating residues into the soil, and choosing the degree of decomposition of residues when they are incorporated.

Soil-air management may also be attacked from the land-use side. Where it is difficult to increase aeration, crops can be grown that have either lower oxygen requirements or shallow-root systems and thus live in the best-aerated part of the soil profile.

REFERENCES

Bertrand, A. R., and H. Kohnke: Subsoil Conditions and Their Effects on Oxygen Supply and the Growth of Corn Roots, *Soil Sci. Soc. Am. Proc.*, vol. 21, pp. 135–140, 1957.

Domby, C. W., and H. Kohnke: The Influence of Soil Crusts on Gaseous Diffusion, *Soil Sci. Soc. Am. Proc.*, vol. 20, pp. 1–5, 1956.

Evans, D. D.: Gas Movement, "Methods of Soil Analysis," pt. 1, pp. 319–330, Am. Soc. of Agron., Inc., 1965.

Janert, H.: Die Durchlüftbarkeit des Bodens, *Trans. 6th Comm., Intern. Soc. Soil Sci. B.*, pp. 468–473, 1937.

Lawton, K.: The Influence of Soil Aeration on the Growth and Absorption of Nutrients by Corn Plants, *Soil Sci. Soc. Am. Proc.*, vol. 10, pp. 263–268, 1945.

Lemon, E. R., and A. E. Erickson: Measurement of Oxygen Diffusion in the Soil with a Platinum Microelectrode, *Soil Sci. Soc. Am. Proc.*, vol. 16, pp. 160–163, 1952.

Letey, J. and L. H. Stolzy: Measurement of Oxygen Diffusion Rates with the Platinum Microelectrode, I, Theory and Equipment, *Hilgardia*, vol. 35, pp. 545–554, 1964.

Nietsch, W. v.: Der Porengehalt des Ackerbodens: Messverfahren und ihre Brauchbarkeit, *Bodenk. Pflanzenern.*, vol. 1, pp. 110–115, 1936.

Raney, W. A.: Field Measurement of Oxygen Diffusion through Soil, *Soil Sci. Soc. Am. Proc.*, vol. 14, pp. 61–65, 1949.

Robinson, F. E.: A Diffusion Chamber for Studying Soil Atmosphere, *Soil Sci.*, vol. 83, pp. 465–469, 1957.

Russell, M. B.: A Simplified Air-picnometer for Field Use, *Soil Sci. Soc. Am. Proc.*, vol. 14, pp. 73–76, 1949.

Stolzy, L. H., and J. Letey: Correlation of Plant Response to Oxygen Diffusion Rates, *Hilgardia*, vol. 35, no. 20, pp. 567–576, 1964.

Van Bavel, C. H. M.: Composition of Soil Atmosphere, "Methods of Soil Analysis," pt. 1, pp. 315–318, Am. Soc. of Agron., Inc., 1965.

Wilde, S. A., and G. K. Voigt: "Analysis of Soils and Plants for Foresters and Horticulturists," J. W. Edwards, Publisher, Incorporated, Ann Arbor, Mich., 1955.

8 SOIL TEMPERATURE

OUR INTEREST IN SOIL TEMPERATURE

Temperature is an extremely important property of the soil. It affects plant growth directly and it also influences moisture, aeration, structure, microbial and enzyme activity, the decomposition of plant residues, and the availability of plant nutrients. For these reasons it deserves our closest attention.

Soil temperature is one of the important growth factors of plants, just as water, air, or nutrients. Seeds, plant roots, and microbes live in the soil, and their life processes are directly affected by the temperature of the soil. Frequently we do not realize the importance of the temperature of the soil upon plant growth, because it is similar to the temperature of the air and we ascribe any effect upon plant growth to air temperature. But in some cases great divergences exist. A wet soil in the spring may remain cold for quite some time after the atmosphere has become warm. Nitrification is inhibited and the low temperature slows down the intake of water by the roots. It is rare, however, that extreme heat of the soil as such injures plant growth directly.

Of great importance to plant growth is the effect soil temperature exerts upon soil moisture. Soil aeration is affected by temperature and moisture-content differences. The effect of temperature upon soil structure is through its effect upon plant growth and moisture changes and through freezing and thawing. Soil temperature, either by mere raising or lowering or by freezing the soil water, has a pronounced effect upon the decomposition of organic and mineral components of the soil, the resulting release of plant nutrient elements, as well as on clay formation. The rate of chemical reactions doubles with every rise of 10° in temperature.

The Great Soil groups of the world owe their properties to the combined influence of temperature and moisture on soil-forming material (Fig. 2-16).

THERMAL CONCEPTS AND UNITS

HEAT

Heat is the kinetic energy of the random motion of the ultimate particles of which material bodies are composed.

TEMPERATURE

The term "temperature" is used to refer to a particular level or degree of molecular activity. Temperature is the intensity of heat or the level of heat. In the "Handbook of Chemistry and Physics" (published by the Chemical Rubber Company) temperature is defined as "the condition of a body which determines the transfer of heat to or from other bodies." Temperature is measured in degrees. The centigrade scale described in 1742 by Anders Celsius, a Swedish astronomer, is the most practical one and is used in all countries of the world except in some of the English-speaking ones. A centigrade degree is one-hundredth of the difference between the temperature of melting ice and that of boiling water under standard atmospheric pressure (760 mm of mercury). The centigrade scale starts at the temperature of melting ice. The same temperature interval is used also to express absolute temperature (degrees Kelvin) except that the scale starts at −273.18°C, which is absolute zero.

About 1724 a German glass blower, Fahrenheit, working in Danzig, developed a temperature graduation system in which he used the coldest temperature of an ammonium chloride–ice-water mixture as the zero point and blood heat as 100°. This system has so far persisted in most English-speaking countries.

THERMAL CAPACITY

The amount of heat existing in a body is called its *thermal capacity* or *heat capacity*. The thermal capacity of a substance can be defined as the amount of heat required to change the temperature of a given mass of that substance by a certain amount. The unit of heat capacity is the gram calorie, the amount of heat required to raise one gram of water from 15 to 16°C. The specific heat is the heat capacity of a substance in relation to that of water. The specific heat of water therefore is 1.00 cal/g. That of many soil-forming minerals is near 0.2 cal/g. Practically all substances have heat capacities smaller than that of water.

The quantity of heat required to produce a given temperature change depends upon the mass and the nature of the object being heated. Expressed as an equation

$$H = sm\Delta t$$

H = amount of heat in calories, ML^2T^{-2}

s = thermal capacity, $L^2T^{-2}\theta^{-1}$

m = mass, M

t = temperature, θ

THERMAL CONDUCTIVITY

Thermal conductivity is the ability of a substance to transfer heat from molecule to molecule. For this reason it is sometimes called *molecular conductivity*. It is defined as the quantity of heat transmitted through a substance per unit cross section, per unit of temperature gradient. It is expressed in calorie centimeters per second per square centimeter per degree-centigrade temperature gradient. The dimensions are $MLT^{-3}\theta^{-1}$ or cal/sec-cm-°C.

Actual heat conductance depends on thermal conductivity and on the heat gradient.

THERMAL DIFFUSIVITY

Thermal diffusivity is an expression of the rate at which a substance heats up as the result of a thermal gradient. It is the rate of change of temperature with time. It is proportional to thermal conductivity and inversely proportional to heat capacity on a volume basis.

$$\text{Thermal diffusivity} = \frac{\text{thermal conductivity}}{\text{volumetric heat capacity}} = \frac{\text{thermal conductivity}}{\text{specific heat} \times \text{density}}$$

Thermal diffusivity is also called *coefficient of thermal diffusion* or *thermometric conductivity*. Its dimensions are

$$\frac{MLT^{-3}\theta^{-1}}{L^2T^{-2}\theta^{-1} \times ML^{-3}} = L^2T^{-1}$$

Whereas the thermal diffusivity expresses the rate at which a body warms up under a given heat gradient, its reciprocal indicates the ability of the body to retain heat. Therefore it can be called *thermal retentivity*.

$$\text{Thermal retentivity} = \frac{1}{\text{thermal diffusivity}}$$

The dimensions of retentivity are $L^{-2}T$.

RADIATION

Thermal radiation is the transfer of heat energy across space without a carrier medium. Radiation from a very hot body, e.g., the sun, is in short waves. Such short waves—from 0.3 to 2.2 microns—are the most efficient for heat radiation. Soil and atmosphere radiate in long waves—6.8 to 100 microns. These have much less penetrating power. An ordinary glass pane can stop them ("greenhouse effect"). Wavelengths between 2.2 and 6.8 microns are inefficient for heat radiation. Radiation is measured as heat energy per area per unit of time (cal/cm^2-min). The dimensions are

$$\frac{ML^2T^{-2}}{L^2 \times T} = MT^{-3}$$

FACTORS AFFECTING SOIL TEMPERATURE

The temperature of the soil is determined by the interaction of numerous factors. Ultimately all soil heat comes from two sources: radiation from the sun and sky and conduction from the interior of the earth. This latter source is very minor in importance. Both external (environmental) and internal (soil) factors contribute in bringing about changes of soil temperatures.

ENVIRONMENTAL FACTORS

Solar radiation The amount of heat from the sun that reaches the earth is 2.0 g-cal/cm^2-min (or 2.0 langleys/min). The amount of radiation that is actually received by the soil surface is much less. It depends on

1. The angle with which the soil faces the sun, due to latitude, season, time of day, steepness and direction of slope, and the altitude of the location.
2. The insulation by air, water vapor, clouds, dust, smog, snow, plants, or mulch.

In the temperate zone, between 100 and 800 langleys of radiation are received at the earth's surface per day. Five-hundred eighty langleys of energy are required to evaporate a layer of water 1 cm thick.

Of course only a part of the total radiation is available to supply energy for evaporation and transpiration. The rest heats up the soil, is used in photosynthesis, or is reradiated into the sky.

Solar radiation occurs as short-wave radiation, the wavelengths ranging from 0.3 to 5.0 microns.

Radiation from the sky Radiation from the sky contributes a relatively large amount of heat to the soil in areas where the sun's rays have to penetrate the earth's atmosphere very obliquely. Under such conditions much of the sun's energy is absorbed by the atmosphere and is radiated in all directions. In tropical countries the sun's rays pass through the atmosphere more nearly vertically and lose little of their energy. Therefore the proportion of radiation from the sky under such conditions is small.

Conduction of heat from the atmosphere Since the conduction of heat through air is small, it can have a substantial effect upon soil temperature only by contact. This means that air convection or wind is necessary to heat up the soil by conduction from the atmosphere. The classical examples of this phenomenon are the föhn of the Alps and the chinook of the American Northwest. These winds are so warm that within a few minutes they can melt a layer of snow and completely dry up the soil surface.

Condensation Condensation is an exothermic process. Whenever water vapor from the atmosphere or from other soil depths condenses in the soil, it heats up noticeably. Under such conditions, increases of 5°C and more in soil temperature have been noted. In a similar way, freezing of water generates heat.

Evaporation Evaporation, an endothermic process, works in the other direction. The greater the rate of evaporation, the more the soil is cooled down. Moist soils seldom are very hot. Also thawing of ice absorbs heat.

Rainfall Depending on its temperature, rainfall can cool or warm the soil.

Insulation The soil can be insulated from the environmental temperature factors by a plant cover, mulch, snow, and also clouds and fog. Insulation serves in all cases to maintain a more uniform soil temperature. During the summer, insulated soil is cooler than soil that is directly exposed to "the elements," while in winter the situation is reversed.

Vegetation Transpiration of water, reflection of incident radiation (back radiation), and energy used for photosynthesis by plants tend to decrease the temperature of the microclimate and indirectly of the soil. As previously mentioned, a plant cover serves as insulation and consequently tends to smooth out soil temperature fluctuations.

SOIL FACTORS

Thermal capacity The specific heat of dry mineral soil is near 0.2 cal/g. This means that 1 cc of dry soil consisting half of solids and half of pore space would have a heat capacity of $0.5 \times 2.65 \times 0.2 = 0.265$ cal/cc (or 0.25 cal/cc as a rough average) since the heat capacity of air is negligibly small. Such a soil with all its pores filled with water would have a specific heat of $0.265 + (0.5 \times 1.0) = 0.765$ cal/cc. If only half of the pore space (one-quarter of the total volume) were filled with water, its specific heat would be $0.25 + (0.25 \times 1.0) = 0.50$ cal/cc (Fig. 8-1). In case the water in the soil is frozen, the heat capacity is greatly reduced, because ice has a specific heat of only 0.5 cal/g.

The specific heat of muck or peat is larger than that of mineral soil if equal weights are considered, but on equal-volume basis, the differences are small. As organic soils usually have a large percentage of pore space, they have a very high heat capacity when they are water saturated (about 0.9 cal/cc).

The specific heats of some substances are given in Table 8-1 on weight basis and on volume basis.

Thermal conductivity and diffusivity The thermal conductivity of soil-forming materials and most soil particles lies around 0.005 thermal conductivity units; that of air is about 100 times smaller, while that of water is about one-fifth that of the soil-forming minerals. For this reason, dry, loosely packed soil has a very low thermal conductivity

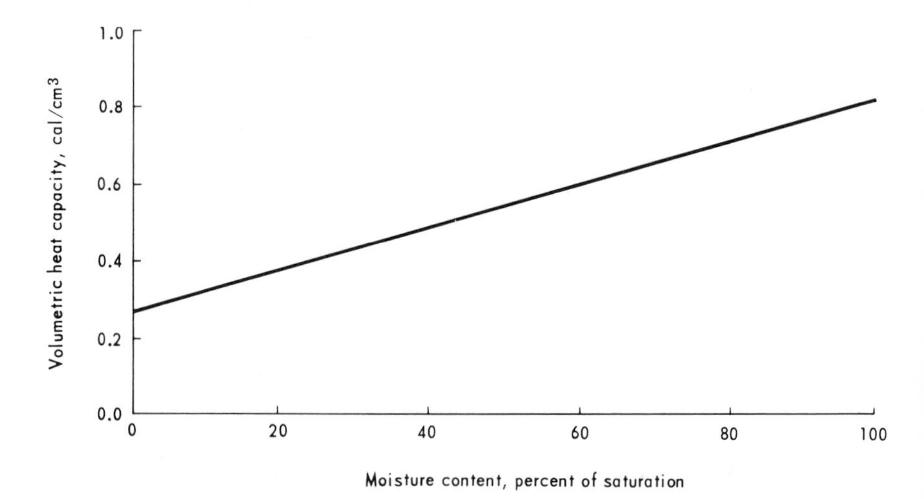

Fig. 8-1 Effect of moisture content on the heat capacity of soil: Assuming 0.2 cal/g of soil, 2.65 g/cc density, 50 percent pore space.

Table 8-1 Specific heat of some substances

Material	*Specific heat, cal/g*	*Volumetric heat capacity, cal/cc*
Humus	0.4	0.56
Water	1.0	1.0
Ice	0.5	0.46
Air	0.25	0.0003
Clay	0.22	0.5
Quartz	0.19	0.5
Mica	0.21	0.59
Granite	0.19	0.5
$CaCO_3$	0.20	0.54
Fe_2O_3	0.15	0.75
Chalk	0.21	0.46
Wood	0.42	0.38

(0.0003 to 0.0005 cal/sec-cm-°C) because the contact areas between the individual particles are small and the conductivity of air is minute.

An increase of moisture content in the dry range, from oven dry to air dry, increases the thermal conductivity of a soil only very little. As soon as the water starts to form bridges from particle to particle, thermal conductivity increases markedly. This happens when the soil begins to look moist (about 30 atmospheres) (Fig. 8-2, Nakshabandi and Kohnke, 1965).

At a soil-moisture tension of around 10 atmospheres, the thermal conductivity is similar to that of water. When the soil becomes wetter, its thermal conductivity rises considerably beyond this level.

In no case does wet soil conduct heat as fast as solid rock, because the heat conductivity of water is considerably lower than that of the minerals. The thermal conductivity of soil organic matter is about one-half of that of the mineral components. Compared to the metals, the thermal conductivity of all soil components—mineral and organic particles, water and air—is very low. The specific thermal conductivities for metals are: silver = 1.0, copper = 1.0, aluminum = 0.50, and iron = 0.16.

It is fortunate that, as heat conductivity increases—with moisture content—heat capacity also increases (Fig. 8-3). This means that the diffusivity changes much less with moisture changes than does the conductivity. If it were the other way, soils might easily suffer from temperature extremes.

Since thermal diffusivity is proportional to thermal conductivity and inversely proportional to the specific heat per unit volume, the increase of thermal diffusivity of soil with increase of moisture content is

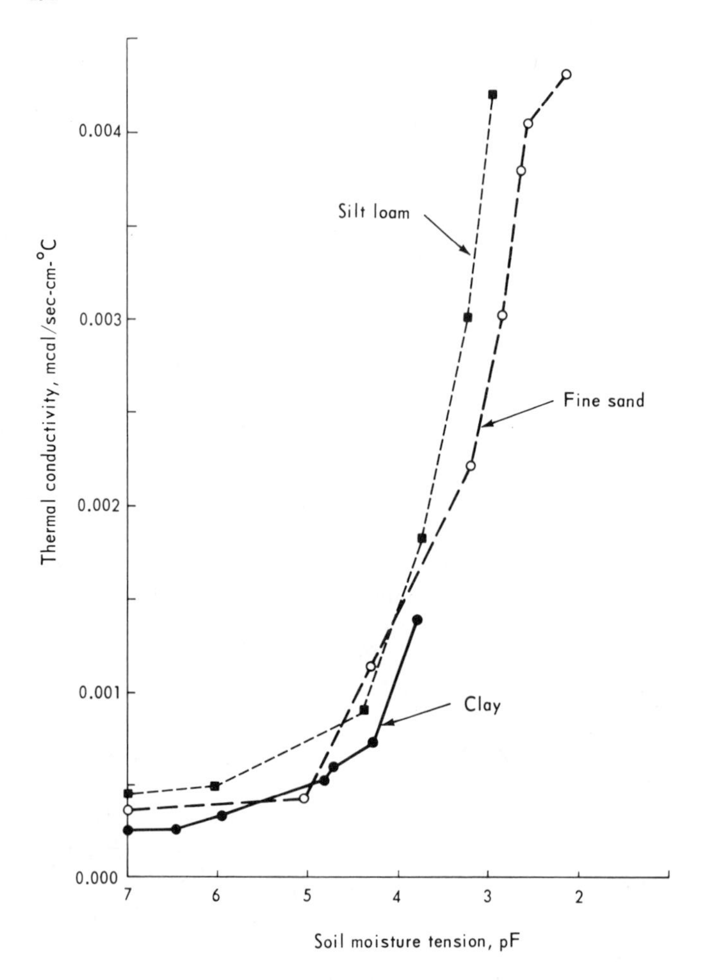

Fig. 8-2 Thermal conductivity of soils related to moisture tension.

more gentle than the increase of thermal conductivity. As a matter of fact at approximately 1 atmosphere tension thermal diffusivity reaches a maximum and rapidly decreases as the soil gets wetter. The reason for this is that, in this range, the specific heat increases faster than the thermal conductivity.

In considering the effects of heat conductivity on the temperature of a specific part of the soil, it must be remembered that heat may flow up or down in the profile, depending on the temperature gradient.

Soil that has high diffusivity greatly affects the air temperature near the ground. Soils of low diffusivity—wet mucks or very loose, crumbly soils—allow air temperatures to change violently.

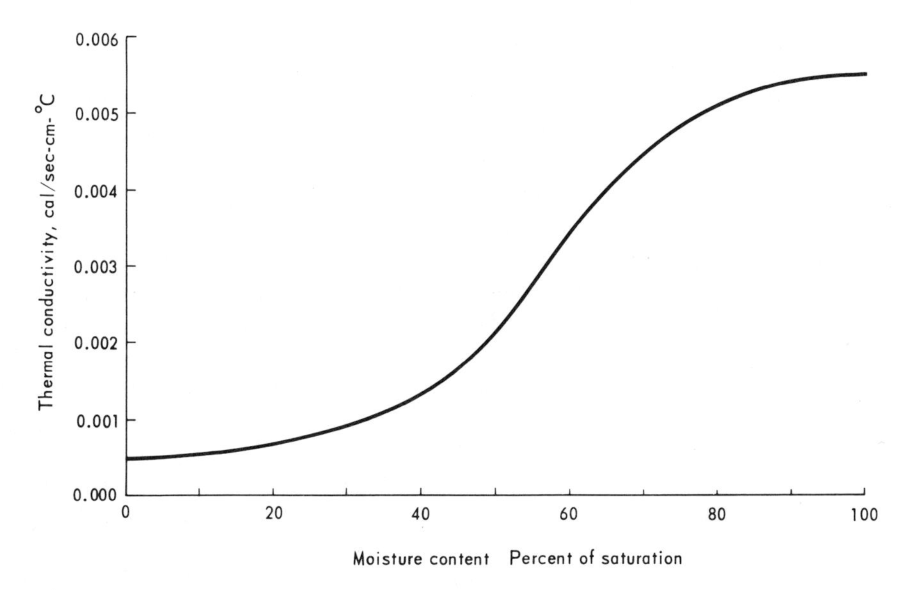

Fig. 8-3 Effect of moisture content on the thermal conductivity of a soil of medium texture.

Table 8-2 Thermal conductivities*

Air	0.5×10^{-4}
Dry soil	$3\text{–}5 \times 10^{-4}$†
Snow	3.9×10^{-4}
Water	$13\text{–}14 \times 10^{-4}$‡
Building brick	14×10^{-4}
Glass	$10\text{–}25 \times 10^{-4}$
Moist soil	$20\text{–}40 \times 10^{-4}$
Ice	$20\text{–}50 \times 10^{-4}$
Soil-forming minerals	$30\text{–}300 \times 10^{-4}$
Iron	1600×10^{-4}
Aluminum	5000×10^{-4}
Silver	$10{,}000 \times 10^{-4}$ $(=1.0)$

* In calorie centimeters per second per square centimeter per degree-centigrade change in temperature gradient.

† Depending largely on compaction.

‡ Depending on temperature.

Biologic activity Biologic activity evolves heat, and the greater this activity in the soil, the more it is heated up. Soil high in organic matter, mineral nutrients, air, and sufficient moisture can heat up several degrees over biologically inactive soils nearby. The so-called cold frames are an example of the utilization of this source of heat.

Radiation from the soil Radiation of heat from the soil to the atmosphere occurs continuously. The higher the temperature of the soil, the greater is the radiation. Radiation through vacuum is greater than through air, since air absorbs part of the radiated energy. Fog, clouds, water vapor, and mulch reduce heat radiation from the soil. Heat radiation varies as the fourth power of the absolute temperature of the radiating object. Radiation at high temperatures (e.g., from the sun) is mostly in short (0.3 to 2.2 microns) electromagnetic waves (visible spectrum), whereas that from objects of lower temperature (e.g., the heated soil) is in long (6.8 to 100 microns) electromagnetic waves. The latter have very little ability to penetrate water vapor, air, and glass. It is for this reason that soils remain warm during a cloudy night or in a greenhouse.

Color has a considerable effect on the reflection of incoming short-wave radiation. The darker the color, the smaller the fraction of the incoming radiation that is reflected. This ratio is called *albedo.*

$$\text{Albedo} = \frac{\text{reflected energy}}{\text{incident energy}}$$

The larger the albedo, the cooler is the soil—other things being equal.

Dark soils and moist soils reflect less than light-colored and dry soils. A soil with a rough surface absorbs more solar radiation than one with a smooth surface.

Structure, texture, and moisture Soils in a compacted condition have a greater thermal conductivity than the same soils in a more loose condition. This is because the mineral particles conduct heat much more than air. Also less water is needed in a compacted soil to bridge the gaps between particles or aggregates.

A soil in its natural structure has higher conductivity than when it is recently disturbed because of the more intimate contact in the natural soil between the individual soil particles. Conductivity increases with bulk density. Soil of platy and blocky structure has a higher thermal conductivity than soil of granular structure. Mineral soil has higher conductivity than organic soil. The differences of thermal conductivity due to structure and texture are small compared with those due to moisture changes.

It has been discussed previously that moisture has an effect on heat capacity and heat conductivity. Moisture at the soil surface, through evaporation, cools the soil. Much of the solar radiation energy is used for evaporation. A moist soil will therefore not heat up as much as a dry one.

Moist soil, being a better conductor of heat than dry soil, is more uniform in temperature throughout its depth.

Thermal data for the most important soil components are summarized in Table 8-3. The concept of thermal retentivity is introduced in this table to express the different abilities of various substances to maintain their temperature when exposed to a temperature gradient. Thermal retentivity is the reciprocal of thermal diffusivity. The high thermal retentivity of water is the reason why the temperature of saturated soils changes only slowly.

Soluble salts The concentration of soluble salts in the soil affects evaporation and indirectly soil temperature. Through their effect on soil fertility, soluble salts affect biologic activity and indirectly soil temperature. A high concentration as well as an extremely low concentration suppresses biologic activity.

SOIL–TEMPERATURE FLUCTUATIONS

Temperature is one of the most dynamic properties of soil. It is subject to daily and seasonal changes. These fluctuations affect other soil conditions and plant life. It is important to study both the amounts as well as the depth of the fluctuations. The factors that affect soil-temperature fluctuations are the same that affect soil temperature as such.

Table 8-3 Thermal data for soil components—approximate averages

		Heat capacity				
Soil components	*Density, g/cc*	*Weight basis, cal/g-°C*	*Volume basis, cal/cc-°C*	*Thermal conductivity, cal/cm-sec-°C*	*Thermal diffusivity, cm^2/sec*	*Thermal retentivity, cm^{-2} sec*
Soil-forming minerals	2.65	0.20	0.53	0.010	0.019	52
Water (liquid)	1.00	1.00	1.00	0.0014	0.0014	713
Ice	0.917	0.505	0.474	0.005	0.0105	95
Dry air (motionless)	0.0012	0.25	0.0003	0.00005	0.167	6

DAILY FLUCTUATIONS

Daily fluctuations are great only in the surface soil. At a depth of 30 cm they seldom exceed 3°C, at 60 cm they may reach about 1°C, while at 1 m diurnal temperature fluctuations are practically nil. These figures refer to summer conditions in the American Midwest where fluctuations are relatively large. In winter these fluctuations are considerably smaller as freezing of soil water and melting of ice in the soil have an equalizing effect upon temperature due to the energy used up and released in these processes. Besides, the amount of energy supplied by the sun during a winter day is much smaller than during a summer day. A snow cover, acting as insulation, also tends to stabilize soil temperatures. In winter (in many areas) soils are generally wetter than in summer and the increased heat capacity results in smaller temperature fluctuations.

During a clear sunny day, a bare soil surface is usually warmer than the air temperature. The time of the peak temperature of both the soil surface and the air coincide approximately. The farther down we measure the temperature in the ground, the greater is the lag of the soil-temperature maxima and minima behind those at the surface, since it takes time for the heat to penetrate the soil. The amount of this lag depends on the nature of the soil, its structure, and its moisture content, all of which determine its thermal diffusivity. The rate of penetration of a heat wave within the soil varies around 3 hours/10 cm of soil depth (or $\frac{3}{4}$ hours/in.).

The cooling period of the daily cycle of soil-surface temperature is about twice as long as the warming period. This is because the rate of heat gains by incoming (short-wave) radiation exceeds the continuous heat loss by out-going (long-wave) radiation normally only from sunup to 1 or 2 hours after noon (Fig. 8-4).

Large daily temperature fluctuations in the soil profile are brought about by high insolation, low albedo, dry surface soil, and high thermal

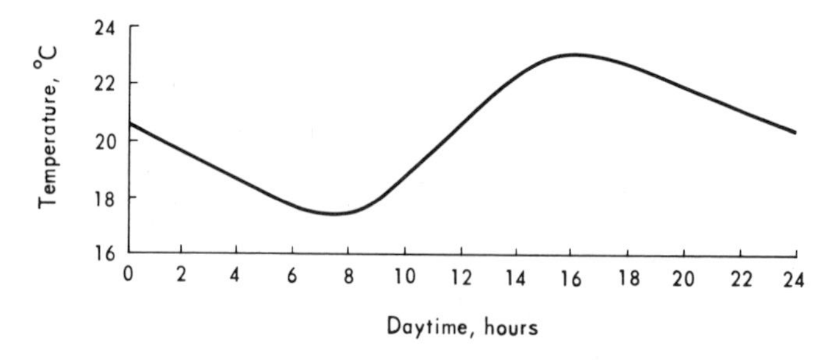

Fig. 8-4 Diurnal temperature fluctuation in a surface soil: Depth, 10 cm; cover, grass.

diffusivity. If thermal diffusivity is low but the other properties are as described, the surface soil heats up greatly without causing major temperature changes in the body of the soil. This situation can exist on a recently cultivated field after the surface soil has dried out.

Daily temperature fluctuations are minimized by high water content, clouds, mulch, snow, plant cover, and an aspect away from the sun (north-facing slopes in the Northern Hemisphere).

SEASONAL FLUCTUATIONS

Seasonal temperature fluctuations extend much deeper into the ground. They may reach to a depth of 10 m or even more. The greatest depths of seasonal fluctuations are found in continental climates of the temperate zone because the difference between summer and winter surface temperatures is largest. Tropical and oceanic regions have less differences between summer and winter temperatures, and in the arctic zones ice acts as a temperature stabilizer.

Under woods or other dense plant cover, both the daily and the seasonal temperature oscillations are smaller and extend less deeply than where no or only sparse vegetation exists.

It is interesting that at the depth where the temperature is constant throughout the year, it is slightly above the average air temperature of the locality. Springs have usually a temperature quite close to this. This is affected by the depth of the source. Shallow springs have a more variable temperature. Springs with very deep sources are hotter than the average local temperature.

The greatest part of the radiation energy from the sun is absorbed by the soil and the plants growing on it. The atmosphere absorbs a much smaller amount. It receives most of its heat through radiation from the earth. For this reason the average annual temperature of the soil, at all depths, is higher than the average annual temperature of the air at the same location. In the United States this difference is about 1°C, except in parts of the northern states and in the higher mountains where snow insulates the soil for extended periods. Here the difference is still greater (Smith *et al.*, 1964).

The depth to which soil freezes depends on the severeness and the duration of the cold spells. Maximum penetration of frost is shown in the following table.

6 cm	Auburn, Alabama
60 cm	Lafayette, Indiana
90 cm	Ottawa, Ontario
Over 150 cm	Scott, Saskatchewan
"Permafrost"	Central and northern Alaska

The season of greatest frost penetration in the American Midwest is late January and February. February is the month of coldest subsoils. February is practically the only month when soil in protected, well-littered woodland freezes in Indiana.

Soil-temperature fluctuations are greatly affected by topography. Slopes facing the sun have greater temperature fluctuations, both diurnal and seasonal, than those facing away from it. The greater the steepness of the slope (up to vertical incidence of the sun's rays), the more pronounced is this difference. Because of air drainage, air temperatures are lower on the foot of the slope than at higher points. This can cause a noticeable effect on soil temperatures.

EFFECT OF SOIL TEMPERATURE ON OTHER SOIL CONDITIONS

SOIL MOISTURE

A change in temperature affects vapor pressure and viscosity of the soil water. In a moist soil the relative humidity of the air is between 98 and 100 percent. This means that the vapor pressure of the water under such conditions can change materially only as a result of temperature changes. These changes can cause a vapor pressure gradient between different depths of the soil and result in movement of water in the vapor phase. This movement can take on considerable proportions and be of importance in supplying plant roots with water. It is especially pronounced if one layer of the soil freezes and the vapor pressure is therefore maintained at a constant low level. The vapor is condensed and frozen and therefore does not contribute to a buildup of water vapor pressure. These effects have been discussed in detail in Chap. 5.

A change in viscosity due to temperature changes is of minor effect on infiltration or percolation in most soils, since a rise in temperature is associated with swelling of moist colloids and therefore the greater ease of movement of the water at the higher temperatures is partially compensated by the reduced diameters of the soil pores.

Infiltration can occur even in a frozen soil if sufficient large pores are not filled with ice. This situation usually exists at the beginning of winter when a moist soil freezes. After the soil gets nearly saturated with water and then freezes, its passageways are blocked with ice and infiltration stops. This is one of the reasons that usually not the first but the subsequent large rainstorms in the winter are followed by floods.

SOIL STRUCTURE

Accumulation of water during periods of frost in the upper layers of the soil causes dispersion of soil particles. The same result is obtained—if not so pronounced—by the expansion of water upon freezing. To what extent this is beneficial or deleterious depends upon moisture content, soil texture, the original soil structure, and the rate of freezing, and has been discussed in Chap. 5.

Since soil temperature affects the amount of organic matter in a soil, it is of great indirect importance in establishing soil structure. A similar statement can be made about the type of colloidal clay in the soil. Climates with intermediate temperatures (5 to 15°C annual average) give rise to soils with the highest amounts of aluminosilicate clays. Very hot and very cold climates bring about soils with very small clay contents.

HEAVING

Intermittent freezing and thawing of the soil causes heaving of plant roots or of stakes driven into the ground and of the soil itself. As the water freezes near the surface, beginning in the larger pores, more water comes up from the unfrozen soil farther down—mostly by capillarity, but also in vapor form—and the ice lenses formed in the soil lift the surface soil with everything in it. The downward progress of freezing in the soil is slow because of the heat of fusion created. This heat has to be dissipated at the soil surface. Below the frozen layer the heat of fusion helps to maintain the soil water in liquid condition. The water can therefore move up and continue to increase the ice lens. As the frozen soil at the surface clings tightly to the plant roots, they are pulled up along with the soil. When the ice thaws in daytime and the water seeps away, the soil settles again but leaves the plant root in the position it was while the soil was frozen because the entire root has been pulled out and cannot slip back. This heaving process may be repeated night after night, until the root crown sticks out several centimeters above the ground. Frequently the roots are held so securely in the ground that the force of heaving tears them asunder. The amount and rate of frost heaves depend on the rate of replenishment of moisture below the growing ice. Since high soil-moisture content and frequent changes from thawing to freezing are needed for this process, extensive heaving occurs usually in late winter and especially on soils of high capillary conductivity. A high groundwater table and tight subsoils favor heaving. Silt loams are most liable to heaving, then follow clays, with sands least subject to heaving.

Whenever the ground is insulated with mulches or with densely growing plants, the soil does not freeze and thaw as quickly and as frequently, and heaving does not occur so much. This is the reason that an alfalfa-grass meadow is much safer from heaving than a stand of pure alfalfa. Generally, heaving occurs where there is no complete plant cover of the soil. This is Nature's way of preparing ground for seeds to find a good seedbed.

A special type of heaving is the formation of needle ice. Ice crystals develop parallel to each other and vertically to the soil surface. Usually a thin layer of soil or organic debris is lifted up by the needle ice. The conditions of its formation are a sufficient supply of capillary water at the bottom of the ice crystals and a slow rate of freezing.

MICROBIAL ACTIVITY AND PLANT NUTRIENTS

Soil temperature has a profound influence upon microbial activity. This is very restricted at temperatures below 10°C, the optimum rate of activity of the beneficial soil organisms occurring between 18 and 30°C. Nitrogen-fixing bacteria function best in warm, well-drained soils. At temperatures considerably above 40°C microbes become inactive. Nitrification is dependent upon temperature, the range around 30°C being the optimum. Above 30°C more potassium is released from the nonexchangeable form than at lower temperatures. Low temperatures retard the uptake of potassium by plant roots.

SOIL DEVELOPMENT

Temperature is an important climatic factor in soil development. It has a direct effect on all reactions taking place in the soil. The amount of heat energy available in the soil—which is related to temperature—determines the rate of evaporation of precipitation water. Thus temperature helps to control effective climatic humidity. Even aside from its effects upon moisture, temperature is the determining factor of the occurrence of the Great Soil Groups. A few examples of these are shown in Table 8-4 together with the temperature range in which they occur (Kohnke, Stuff, and Miller, 1968).

Table 8-4 Temperature relations of Great Soil Groups

Average annual temperature, °C	*Great Soil Group*	*Connotation of soils according to the "Comprehensive System"*
Below 0	Tundra	Cryic
0–8	Podzol, gray wooded	Spodic
8–15	Brunizem	Humic mollic
	Gray-brown podzolic	Alfic
12–17	Reddish chestnut	Xeric mollic
	Red and yellow podzolic	Ultic
Above 17	Latosols	Oxic

Table 8-5 Soil classification according to temperature

Mean annual soil temperature, °C	*Soils with 5°C or more difference between mean summer and mean winter temperatures*	*Soils with less than 5°C difference between mean summer and mean winter temperatures*
Less than 8	Frigid	Isofrigid
8–15	Mesic	Isomesic
15–22	Thermic	Isothermic
More than 22	Hyperthermic	Isohyperthermic

In the *Supplement to Soil Classification System* by the Soil Survey Staff (1967) eight soil-temperature classes are presented (see Table 8-5).

Comparing the temperature ranges in Tables 8-4 and 8-5, it appears that the frigid, the mesic, and probably the isohyperthermic classes are represented in both. Since the classification in Table 8-5 has been set up for soil in the United States, it is understandable that Tundra soils are not separated from other soils with a mean annual temperature of less than 8°C. Since most Tundra soils occur in areas with a mean annual temperature below 0°C, it seems they should be placed into a different temperature group than the spodic soils.

A great number of soil groups occur in the intermediate temperature range: about 10 to 20°C. A subdivision of soils into four temperature ranges can obviously serve only as a general classification.

EFFECT OF SOIL TEMPERATURE ON PLANT GROWTH

Soil-temperature requirements of plants vary with the species. It also must be assumed that the soil-temperature requirements of plants vary with the stage of growth. It has been found for corn (Anderson and Kemper, 1964) that yields were approximately doubled by a 10°C rise in temperature above 20°C. At lower temperatures yields decreased by a factor of more than 2 per 10°C decrease in temperature. Similar effects of temperature to growth rates have been found for other plant species. Increasing temperature increases the shoot/root ratio, probably because roots are more efficient in uptake of nutrients and water as temperature is increased (Anderson and Kemper, 1964).

The relation of temperature to plant life may be best portrayed by classifying the effect of temperature on plants in the following manner:

Optimum range: The temperatures under which plants thrive and produce best.

Growth range: The entire range of temperature under which plants can grow. This includes the optimum range.

Survival limits: The maximum and minimum temperatures that can be reached without killing the plants. The actual temperatures depend on the stage of growth and on the duration of exposure to such temperature extremes.

As an example the approximate soil-temperature ranges are given for corn (*Zea mays*, L.) and winter wheat (*Triticum vulgare*, L.).

Range	*Corn, °C*	*Winter wheat, °C*
Optimum range	25–35	15–27
Growth range	10–39	5–35
Survival limits	0–43	−20–43

GERMINATION

In case of germination only soil, not air, temperatures are of importance. Plants have minimum, optimum, and maximum temperatures for germination. For this reason it is wise to place seeds of cultivated crops into the ground only when the temperature is near enough to the optimum to allow fast germination; otherwise the seeds might be attacked by fungi or other pests and might germinate so slowly that weeds having a lower germination temperature will subdue them.

SOIL TEMPERATURES AND AIR TEMPERATURES

After the plants have developed above the ground it becomes difficult to separate the effect of air temperature and that of soil temperature. Generally the fluctuations of temperature in the soil are smaller than those in the air and therefore it can be assumed that air temperatures will be the limiting factor of growth more frequently than soil temperature. However, in spring soil temperatures are generally lower than air temperatures and may restrict microbial life—and thus availability of nutrients—and the ability of the plant roots to take up nutrients and water. In fact, when the soil is frozen and the air warms up, conifers may suffer from "physiologic drought."

AVAILABILITY OF SOIL WATER AND PLANT NUTRIENTS

The free energy of water increases with temperature. This means that, up to a physiologic limit, water becomes more available to plants as the soil warms up. Heating up soil at the wilting point will temporarily return the plants to turgidity.

Low temperatures decrease the availability of plant nutrients, especially of those whose availability depends on microbial activity. Crop roots are stunted and stubby and they branch only little when they grow in very low temperatures. Consequently their ability to absorb water and nutrients is reduced.

DESIRABLE SOIL TEMPERATURES

Generally speaking, intermediate soil temperatures, near the optimum for each given crop, are desirable for crop growth. Extremes should be avoided, if possible, because they would slow down the growth of the crop plants and might favor weeds and disease germs. Soil temperatures should not lag too much behind air temperatures in the spring nor should soil warm up too rapidly. This might increase the dangers of either physiologic drought or of frost damage.

Moderate daily variations in soil temperature are desirable. They aid in soil-moisture distribution through distillation and condensation and in aeration of the soil. It is possible that they are directly beneficial to plant growth.

SOIL-TEMPERATURE MANAGEMENT

Our ability of controlling soil temperature is limited. But since even relatively small changes in temperature can have pronounced effects upon plant growth, soil-temperature management can have significant results. We can modify both the intrinsic soil-temperature factors and some of the environmental factors, although the general climate of a given area still escapes our control.

Changes in moisture content and structure can alter thermal conductivity, heat capacity, and consequently thermal diffusivity and the actual soil temperature. An adjustment of the roughness, the moisture content, and the color of the soil surface modify both absorption of heat as well as radiation and conduction of heat energy away from the soil. The moisture-tension level at the soil surface is an important factor in determining the rate of evaporation or condensation. These, in turn, cool or heat the soil.

One of the most effective means of influencing soil temperature is the use of mulch. Crop residues or paper or plastic have been used successfully for this purpose. A cover of clear or green transparent polyethylene film can greatly increase soil temperature by permitting heat energy to enter the soil but reducing greatly the radiation from the soil to the atmosphere and also by suppressing evaporation. In general, mulch has the tendency to reduce soil-temperature fluctuations.

Vegetation shades the ground and uses radiant energy for the process of photosynthesis and for transpiration. Consequently it keeps the soil relatively cool. An effective method to help to warm up the seedbed in the spring is to ridge the soil and to place the seeds in these ridges. Here the soil dries out more quickly and, due to its sloping surface, receives more radiation from the sun.

MEASURING SOIL TEMPERATURE

PURPOSES

Since soil temperature is an important plant-growth factor and besides influences other plant-growth factors (e.g., availability of water and nutrients), it is necessary to measure actual soil temperatures. This will be necessary at various locations as well as at various depths. Both the locations and the depths depend on the purpose of the study. For basic research, measurements throughout the profile might be needed. If we only want to know the temperature as it affects germination or root growth, the depth of the seed or of the roots should be chosen.

INSTRUMENTS

Instruments for the observation of soil temperature include mercury thermometers (usually in a special protective cover), thermocouples, and thermistors. If it is desired to record temperatures continuously, various types of soil thermographs exist. Some of them work on the principle of expansion of liquids or gases in bourdons (pencil-shaped hollow tubes). But also thermocouple potentials and the resistance of thermistors can be recorded (Taylor and Jackson, 1965). Where freezing of soil water is to be studied, electric gypsum soil-moisture blocks of the Bouyoucos type can well be used, as the resistance in these blocks increases greatly when they are frozen.

The International Meteorological Organization recommends as standard depths for soil-temperature measurements 10, 20, 50, and 100 cm. For microclimatological and plant physiological studies it seems necessary to also measure temperatures at a shallower depth, e.g., 2.5 cm.

Measurements of thermal conductivity require special precautions, especially if the soil contains water. As the soil is heated up, water vaporizes and distills. The heat of vaporization and the change in heat capacity change the system. The temperature gradient required for such a determination causes water movement. The evaporation of water at the warm side and condensation at the cool side cause a fictitious thermal conductivity. In addition the change in moisture content itself changes the conductivity of the soil. In order to minimize these effects and to obtain reliable results, it is necessary to measure thermal conductivity during a very short time. (Patten, 1909; Kersten, 1949.)

REFERENCES

Anderson, W. B., and W. D. Kemper: Corn Growth as Affected by Aggregate Stability, Soil Temperature, and Soil Moisture, *Agron. J.*, vol. 56, pp. 453–456, 1964.

Kersten, M. S.: Thermal Properties of Soils, *University of Minnesota Inst. Tech. Bull.* 28, 1949.

Kohnke, H., R. G. Stuff, and P. A. Miller: Quantitative Relations between Climate and Soil Formation, *Zeitschrift für Pflanzenernährung und Bodenkunde*, vol. 119, pp. 24–33, 1968.

Nakshabandi, G. A., and H. Kohnke: Thermal Conductivity and Diffusivity of Soils as Related to Moisture Tension and Other Physical Properties, *Agr. Meteorol.*, vol. 2, pp. 271–279, 1965.

Patten, H. E.: Heat Transference in Soils, *U.S. Dept. Agr. Bur. Soils. Bull.* 59, 1909.

Smith, G. D., F. Newhall, L. H. Robinson, and D. Swanson: Soil Temperature Regimes, Their Characteristics and Predictability, *U.S. Dept. Agr. Soil Conservation Service* SCS-TP-144, 1964.

Soil Survey Staff: "Supplement to Soil Classification System," 7th Approximation, Soil Conservation Service, U.S. Department of Agriculture, 1967.

Taylor, S. A., and R. D. Jackson: Temperature, "Method of Analysis," pt. 1, no. 9 in the Series Agronomy, pp. 331–344, Am. Soc. of Agron., Inc., 1965.

9 SOIL COLOR

Color is one of the most obvious characteristics of soil and one that is probably more frequently used to describe soil than any other. Soil color has no direct effect on plant growth but an indirect one through its effect on temperature and moisture. Color can be an indicator of the climatic condition under which a soil was developed or of its parent material. In many instances the productive capacity of a soil can be judged from its color.

All this points to the importance of familiarity with the subject of soil color.

COLORS THAT OCCUR IN SOILS

Practically all colors occur in soils. This includes white, red, brown, gray, yellow, and black. Even bluish and greenish tinges occur. Predominantly soil colors are not pure, but mixtures, such as gray, brown, and rust. Pure blue and green are not known to exist in soils. Frequently two or three colors occur in patches; this is called *mottling.*

CAUSE FOR SOIL COLORS

The color of the soil is a composite of the colors of its components. The effect of these components on the color of the composite (the soil) is roughly proportional to their total surface which is equal to their specific surface times their volume percentage in the soil. This means that colloidal material has the greatest impact on soil color. The outstanding examples are iron hydroxides and humus. Humus is black or brown; iron oxides may be red, rust-brown, or yellow depending upon the degree of hydration. Reduced iron is blue-green. Quartz is mostly white. Limestones are white, gray, or sometimes olive-green. Feldspars have different colors, with red predominating. Clays are gray, white, or red. The colors of clays are largely determined by the type and the amount of iron coatings.

Mottling is the result of the solution and removal of some of the soil components—particularly iron and manganese—from the soil during the wet season and their precipitation and deposition when the soil dries out. This is largely due to reduction which brings iron and manganese into solution and oxidation which causes their precipitation. The light-colored patches in the soil are low in iron and manganese, while the dark ones show where iron or manganese have been precipitated. Mottling is not readily reversible; even after a soil is well drained with tiles or ditches it will retain mottling.

Wet and moist soils look darker than dry soils. The reason for this is that the refractive properties of the solid soil components and of the air are very different and that therefore the light that falls on a dry soil is largely reflected. The refractive properties of water and soil particles are sufficiently similar that light penetrates the soil and much less of it is reflected.

SIGNIFICANCE OF SOIL COLOR

Color can serve to tell much about a soil. Generally speaking the darker a soil, the higher is its productivity. This is due to the amount of organic matter present and the leaching of plant nutrients that has taken place. Light color frequently results from the preponderance of quartz, a mineral that has no nutritional value. The sequence of decreasing productivity is black, brown, rust-brown, gray-brown, red, gray, yellow, white. Quite obviously this relation has many exceptions. In "young" soils it is an indication of the parent material. In "mature" soils it is an indication of the climate in which they have developed. This refers both to the macroclimate and to the soil climate. A warm climate brings about red soil colors, especially if the soils are well drained. A light color is frequently the result of leaching of iron from the soil. Along with the iron many of the plant nutrients have been washed out. This accounts for the fact that light-colored soils are often low in productivity. Mottled colors in soils indicate intermittent reduction and oxidation and point to a temporary excess of water and a consequent lack of aeration.

Practically all soil profiles reveal a change in colors from one horizon to the next. These changes are most obvious in mature soils, while both in young and very old soils they are less pronounced. In young soils there has not been sufficient time for much differentiation, while in the very old ones leaching has proceeded to considerable depth and has left only the least soluble components.

The specific color of the horizons makes it possible to recognize erosion. In many fields the eroded spots stand out clearly from the rest of the land.

In the classification of soils, color is frequently very helpful. It is employed both in the earlier systems of classification and in the "Comprehensive System" of the U.S. Department of Agriculture (Soil Survey Staff, 1967). Examples are the Great Soil Groups chernozem (black), sierozem (gray), krasnozem (red), podzol (ash-gray), latosol (red), gray-brown podzolic, red and yellow podzolic, gray wooded soils, andosoils (dark), and chestnut soils. The Comprehensive System uses such formative elements in its nomenclature as "alb" (white), "ochr" (light-colored), "umbr" (dark), "sombr" (dark), as well as the term "chrom" to indicate the presence of pronounced color.

Since colors are good indicators of soil characteristics, they serve well in the study of the genesis of soils and in arriving at conclusions concerning their best use and management. Soil color can be a guide to the climatic soil group, to the parent material, or to the physiographic location. Soils in the humid temperate and the cold zones are predominantly grayish. Red and yellow soils are mostly found in the tropics and subtropics, although some soils derived from limestone can be red even in warm temperate regions. Depression soils are darker than adjacent upland soils because of the higher content of organic matter. Soils derived from basic rocks are generally darker than soils derived from acid rocks. This is due to the stability of calcium humus and the dispersion and consequent easy removal of hydrogen humus.

EFFECT OF COLORS ON OTHER SOIL CONDITIONS

The color of the soil affects other soil conditions through its effect upon radiant energy. Black and dark colors absorb more heat than light colors or white. Therefore dark soils tend to be warmer than light-colored soils when the sun shines or when the atmosphere is warm and the soil is able to absorb energy from it. This greater amount of heat energy available to the soil results in higher rates of evaporation. A dark soil under otherwise identical conditions will therefore dry out faster than a light-colored one. A cover of vegetation or mulch will naturally reduce or even eliminate this difference.

Generalizing, it can be said that the primary effect of soil color is on the heat balance. This affects temperature and moisture of the soil, and indirectly plant growth, microbial activity, and soil structure. Only color of the soil surface can have an effect on other soil conditions. Color that is not exposed is of no significance.

STUDY AND CLASSIFICATION OF SOIL COLORS

We recognize a rather narrow range of electromagnetic waves as light. The wavelength of visible light extends from about 0.3 to 0.75 micron. The effect of the light of the various wavelengths affects the human eye very differently. These different impressions are called *color*. Table 9-1 shows the wavelengths and the corresponding colors.

The color of an object depends upon the kind of light which it is capable of reflecting to the eye. Soils, as most other objects, reflect light of a great variety of wavelengths. It is possible by means of a spectroreflectometer to determine quantitatively the contribution of the various wavelengths to the total light reflected from a soil. The human eye observes such polychromatic light as a definite color impression. We describe the color with such terms as "gray-brown," "yellow," etc. It is possible for us to distinguish a great many colors, but it is another matter to describe them accurately. There are several reasons for this.

Every person has a somewhat different impression of a given color, to say nothing of the people that are actually color blind. Second, the color of an object depends both on its own properties and on the quality of light that it reflects. Third, the color impression we receive depends partially upon the texture of the object. The finer the particles of the same color, the lighter appears the color to us. And in addition, our color vocabulary is very confused. What one person calls "reddish-brown," somebody else may call "brownish-red" or even "red."

Since it is impossible to quantitatively describe color by words alone, several attempts have been made to use mechanical devices for this

Table 9-1 The wavelengths of visible light

Color	*Wavelength, microns*
Purple	0.38–0.45
Blue	0.45–0.49
Green	0.49–0.57
Yellow	0.57–0.60
Orange	0.60–0.62
Red	0.62–0.75

purpose and to standardize color descriptions by various numerical systems. One method that has been used is a disk made up of sections of the basic soil colors: red, yellow, black, and white. The width of these sections can be varied. The disk is spun and the resultant color compared to that of the soil (Shaw, 1935; Rice *et al.*, 1941). The segments are adjusted until the composite color matches that of the soil. The color is then described in terms of percent of the area of the standard colors used in the disk. It has been found that a complete match of the soil color with that of the spinning disk is not possible, probably because of the effect of the difference in surface characteristics and the movement on the reflectance of the light.

A method that is favored by many soil specialists is the direct comparison of the soil with chips of standard colors. Frequently the Munsell system is used (Rice *et al.*, 1941; Pendleton and Nickerson, 1951; Pomerening and Knox, 1962). The three basic factors which are the components of color—hue, value, and chroma—underlie the construction of the Munsell charts.

Hue refers to the dominant spectral color or quality which distinguishes red from yellow, etc.

Value or brilliance expresses apparent lightness as compared to absolute white. It refers to the gradations white to black.

Chroma defines the gradations of purity of color, or the apparent degree of departure from neutral grays or white.

The mechanics of the Munsell color system are illustrated in Figs. 9-1 and 9-2. Chips of the main colors occurring in soils are placed on cardboard charts. Each of these chips is designated by a numerical system that includes value, hue, and chroma. A circular hole between two adjoining, and similar, color chips permits one to put a soil ped next to these chips and compare the colors. After the best match is found the soil color is described by the number of the chip to which it corresponds or the number is obtained by interpolation between two chips.

Attempts have been made to use spectroreflectance to designate soil colors quantitatively (Shields *et al.*, 1966). A beam of light is directed toward a soil sample and at the same time toward a reference object. Frequently a pure white object is used, such as magnesium oxide. The spectroreflectometer records the intensity of the reflectance for the entire visible range of wavelengths, or even beyond. So far this method has not been perfected to the point where it can specify a color in conventional terms. At this time its value lies in the possibility to check whether two soil samples have the identical color and to provide a quantitative measure of soil color that is independent of human eyesight.

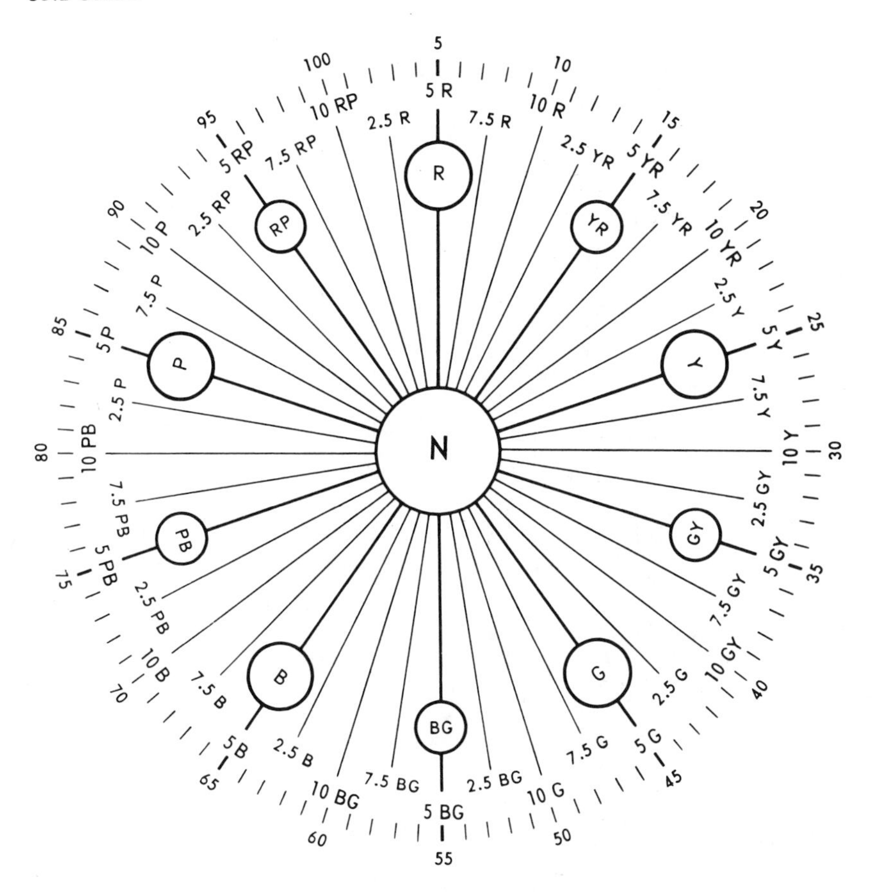

Fig. 9-1 The hue symbols of the Munsell color classification system. (*Courtesy of Munsell Color Company, Inc.*)

Since the color of an object depends on both its own characteristics and the color and intensity of the light it receives, care has to be taken to standardize the source of light when judging soil color by eye. The best light is pure white, as is usually received from the sun around noon. The nearer the sun is to the horizon, the redder is the light. It is not recommended to attempt to determine soil colors within two hours of sunrise or sunset.

Since the color of a soil varies with its moisture content, a decision has to be made at what moisture condition soil colors should be determined. If it is done in the field, the only practical thing is to use moist soil. The moisture content should be high enough to show maximum

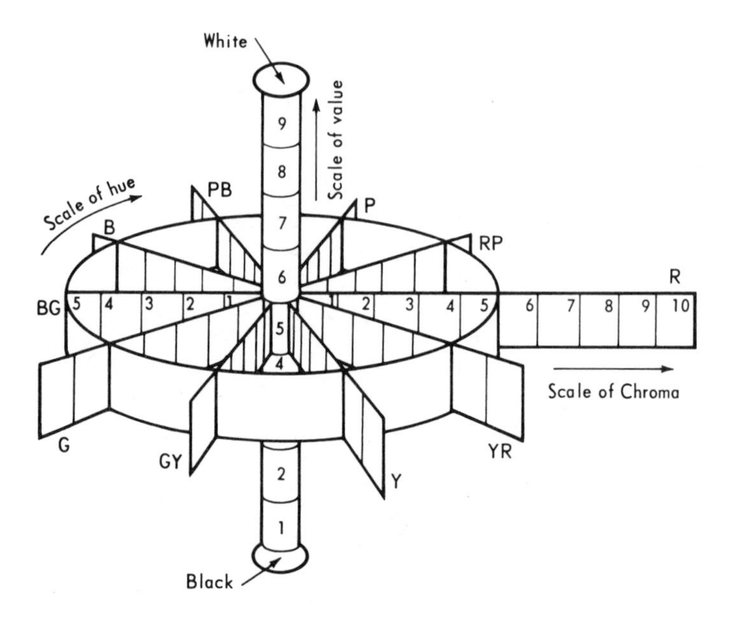

Fig. 9-2 The hue, value, and chroma scales of the Munsell color classification system. (*Courtesy of Munsell Color Company, Inc.*)

darkness. If necessary a small water spray can be used. If the color of the soil is recorded in a dry condition, this should be specified.

MANAGEMENT OF SOIL COLOR

Generally it is not economical to attempt to change the color of the soil surface, if it is done for this purpose alone. In exceptional cases black or white powders or sand or colored plastic sheets have been placed on the soil. The plastic, of course, is used as a mulch to affect temperature and moisture of the soil and its color is incidental. Probably the widest use of a color treatment is the use of crop-residue mulch on top of the ground. By allowing it to decompose early, it can be dark in spring when it is necessary to absorb sunshine energy. Fresh straw, on the other hand, is light, usually lighter than the soil. In some areas of western Canada strips of the fields are cleared of snow so that the sun can directly hit the soil and warm it up instead of being reflected by the whiteness of the snow.

REFERENCES

Pendleton, R. L., and D. Nickerson: Soil Colors and Special Munsell Color Charts, *Soil Sci.*, vol. 71, pp. 35–43, 1951.

Pomerening, J. A., and E. G. Knox: Interpolation of Munsell Soil Color Measurements, *Soil Sci. Soc. Am. Proc.*, vol. 26, pp. 301–302, 1962.

Rice, T. D., D. Nickerson, A. M. O'Neal, and J. Thorp: Preliminary Color Standards and Color Names for Soils, *U.S. Dept. Agr. Misc. Publ.* no. 425, 1941.

Shaw, C. F.: Soil Colors Measured by the Disc Color Analyser, *Trans. Third Intern. Congr. Soil Sci.*, vol. 3, pp. 78–82, 1935.

Shields, J. A., R. J. St. Arnaud, E. A. Paul, and J. S. Clayton: Measurement of Soil Color, *Can. J. Soil Sci.*, vol. 46, pp. 83–90, 1966.

Soil Survey Staff: "Supplement to Soil Classification System," Soil Conservation Service, U.S. Department of Agriculture, 1967.

10 SOIL PHYSICS AS A FACTOR IN SOIL MANAGEMENT

High rates of fertilizer used in the best crop-producing areas of the world are beginning to exclude availability of plant nutrients as a limiting factor. Dense stands of plants with high yielding potential have greater requirements for water and air in the soil and for favorable temperature and structure.

The steady increase in population necessitates also the use of less favored land for agriculture, land in which one or more of the physical plant-growth factors are inadequate. This may be land in regions that are too dry or too cold for best crop production or areas that are difficult to drain or those that suffer from salinity.

All these problems make it necessary to closely analyze the physical conditions of our agricultural soils, if they are to be managed for the benefit both of the individual farmer and of mankind.

PURPOSES OF MANAGING THE PHYSICAL CONDITIONS OF SOILS

FAVORABLE ENVIRONMENT FOR CROPS

The most direct reason for managing physical conditions of soils is to provide a physical environment that is as favorable to crop growth as can be achieved. The factors to be considered are moisture, aeration, temperature, and the resistance to root penetration to the depth normally reached by the given crop. At the same time adjustment of physical conditions has also the purpose of improving chemical and microbiological soil conditions. Both the seedbed and the root bed have to be considered. One of the important purposes is to put the soil in such a condition that the best use can be made of fertilization and other agronomic practices.

SOIL CONSERVATION

The second purpose of managing soil physical conditions is soil conservation. Soils have to be well aggregated and stable to permit maximum infiltration, minimum dispersion, and minimum runoff in order to lose as little of their substance as possible.

Since erosion cannot be avoided completely, it is advisable to establish limits of permissible erosion on the basis of maintenance of soil productivity and of the damage due to excessive sedimentation in the lower reaches of the watershed.

WATER CONSERVATION

Water has become such an important commodity in the life of mankind that its management is the responsibility of the farmer beyond his direct desire to provide the proper amount of water and air for his crops. In general the operations carried out to create a favorable environment for the crops and to conserve soil will at the same time result in the conservation of water. It may be well to state that *conservation* of water means the management of water in such a way that it will do the most good to all the people of an area. It is the intelligent, economic use of water. In farming practice this is accomplished by providing a high infiltration rate that reduces surface runoff and evaporation, increases the water available to crops, and adds to the store of groundwater. In this way water that reaches the streams will be relatively pure and free of sediment.

CROP REQUIREMENTS FOR SOIL PHYSICAL CONDITIONS

These depend on the plants to be grown, the climate of the locality, and indirectly on the economic conditions. Conversely the species of crops to be grown depend on the potentialities of the soil, the climate, and the economic and social conditions. Most of the cultivated plants are mesophytes. They need a constant supply of water at low tension (less than 1 atmosphere) for best growth and highest production, provided the oxygen supply is sufficient.

The requirements for these two components vary widely from species to species due to differences in growth habit and physiologic behavior of the roots. They vary for a given plant with the stage of growth and with temperature.

Some crops require much water and get along on a restricted oxygen supply; others can withstand dry periods without serious damage but suffer quickly when the soil air is exhausted. Some plants can stand low temperatures without ill effects; others die if the temperature is near the freezing point. The optimum soil-temperature range varies for the different crops. It is difficult to separate the plant requirements for soil temperature from those for air temperature. Although the annual averages are nearly the same, soil temperatures usually differ greatly from the ambient air temperatures. For instance so-called cold soils stay cold in the spring much longer than neighboring warm soils. Soil temperature has naturally also a profound effect on the growth of plant disease germs. At certain cool temperatures some pathogenic fungi thrive, while the crop plants grow slowly and easily succumb to the inroads of the disease.

The requirements of crops for penetrability of the soil differ also. Shallow-rooted plants are not handicapped by a hard pan that might greatly stunt alfalfa or sugar beets. For maximum success in crop production a full understanding of the physiologic requirements of the individual plant species is needed and these have to be matched with soils of corresponding characteristics.

Since the environmental requirements vary greatly from crop to crop, it is impossible to set up specifications for the physical conditions of agricultural soils that would be generally applicable. Soil texture, topography, and climate frequently preclude the attainment of such ideals. The desirable conditions of moisture, structure, aeration, and temperature have been discussed in the respective chapters. It is hoped that this information will make it possible to determine the best management of soil physical conditions for any given set of circumstances.

ADJUSTMENT OF LAND USE TO THE PHYSICAL CONDITIONS OF THE SOIL

As our ability to change the physical soil conditions to suit some desired crop is limited, it is wise to adjust as far as feasible the land use to the physical conditions of the soil as well as to the climate.

TEXTURE

On extremely sandy soil that cannot be irrigated it may only be possible to grow pines. If the sand has a greater admixture of finer textural fractions it becomes possible to grow rye. On sandy loam wheat and clover can be grown successfully, and on loams, silt loams, and silty clay loams corn, soybeans, beets, and other crops with higher chemical and physical requirements are in place. Soils with a very high content of aluminosilicate clays do not lend themselves well to tillage. They may be better adapted to pasture crops.

MOISTURE

No crop can produce high yields on soils that are dry (drier than the wilting point) during much of the growing season. But sorghum, the prairie grasses, and some other species suffer less from occasional dry spells than, for instance, corn and Kentucky bluegrass. It is necessary therefore to select such drought-resistant crops in dry regions, if irrigation is impossible. Also the distribution of moisture in the profile is important. A soil of relatively low moisture content is satisfactory for deep-rooting plants like alfalfa and sweet clover provided there is a good moisture supply in the lower strata. Where soil moisture in the summer is not abundant it may be advantageous to grow winter annuals (e.g., winter wheat) to make use of the moisture in the fall and in the spring. Where the moisture of one year does not suffice for a crop, farmers allow the rainwater of two seasons to accumulate in the soil by fallowing the land every other year ("summer fallow"). Crops of high water requirements like corn, sugar beets, and soybeans can only be grown successfully on land with adequate soil moisture. Sometimes reducing the stand can adjust the water needs of the crop to the water supply.

Some plants do well under conditions of abundant soil moisture and restricted aeration, as for instance alsike clover, red top, and reed canary grass.

AERATION

Most common crop plants have a fairly high oxygen requirement. This is particularly true for crops with a deep-rooting habit, e.g., sugar beets and alfalfa. Soils with limited aeration, that are not readily improved, lend themselves only to crops with small oxygen requirements or such that can conduct atmospheric oxygen from the leaves to the roots. Rice is a case in point. Most other cultivated plants do poorly under such conditions. Sorghum is an example of a crop that can survive occasional flooding without ill effects.

TEMPERATURE

Soil-temperature ranges are so similar to those of the air that the predominant crop adjustments are made for air temperature rather than for soil temperature. But "cold" soils actually delay the warming up of the soil in the spring and may require the use of crops with shorter growing periods.

SOIL STRUCTURE

The only direct effect of soil structure on plant growth is the resistance to root penetration. The structure of the surface soil and the structure changes in the profile have to be taken into account in the selection of crops. A dense soil, for instance, may require the use of either shallow-rooted plants or of crops with deep and powerful roots that can overcome the resistance. Indirectly soil structure affects crop growth through moisture, temperature, aeration, and the degree of root-soil contact.

COLOR

No adjustments are necessary for soil color, since it does not affect plant growth directly.

It goes without saying that land use has to be adjusted to other factors besides soil physical conditions. Such factors include climate, topography, groundwater level, soil chemical conditions, the economic situation, and the status of agricultural technology.

THE AIMS OF MANAGING THE PHYSICAL CONDITIONS OF SOIL

While the preparation of the ground for maximum production of food and fiber is the main purpose of soil management by the farmer, soil and water conservation are inseparably connected with it. Fortunately, in actual practice the physical conditions of the soil that are best for the crop are essentially also best for soil and water conservation.

It is impossible to set up specifications of soil physical conditions that fit every crop and every climate. Nevertheless general guidelines will be given that apply under many situations.

WATER

Sufficient water at a tension lower than 14 atmospheres, and preferably lower than 6 atmospheres, must be available in the root zone to supply the plants for a period longer than an anticipated drought.

The lower this tension is, the more favorable is the water situation. To achieve this, the soil should have a high infiltration rate but a high retentivity for water in the lower horizons. A deep penetrability for roots is desirable to allow the plants to forage for water in a large volume of soil. It is advantageous to permit the moisture content of the soil to fluctuate, as drying and wetting help to improve aggregation and aggregate stability physicochemically and through microbial action.

AIR

The replacement of soil air should be sufficient to supply at least 30×10^{-8} g O_2/cm^2-min by diffusion to a depth in the profile where plant roots are expected to grow.

Shortly after seeding this depth is limited by the location of the seed, while in periods of drought this may represent 1 m or more. The need for oxygen is greatest in summer during the most active growth period. At this time the oxygen concentration in the upper 20 cm of the soil should not drop materially below 12 percent.

Another aim of soil-air management is to keep the air always between 99 and 100 percent saturated with water vapor. Since the permanent wilting percentage corresponds to 99 percent relative humidity, a water vapor saturation less than this figure in part of the root zone results in restricted growth or even wilting of the crops.

STRUCTURE

To achieve simultaneously the aims of having ample supplies of water and of oxygen in the soil represents one of the most difficult tasks of soil management. The space taken up by water cannot be taken up by air. This problem leads us to the consideration of the aim of soil-structure management. While climate, slope, groundwater situation, and the crop to be grown modify the specifications for the desired soil structure, the following guidelines have rather general application. The surface soil, down to about 20 or 30 cm, should be rich in noncapillary pores so that water can easily enter (high infiltration rates) and the aggregates should be water stable. Such a condition is called *good tilth.* Soil in good tilth will soon be drained of excess water and will have enough pore space available for air to enter. The nearer to the surface, the less desirable is the presence of a large percentage of capillary pores because they would tend to replenish the moisture at the surface, making it liable to excessive evaporation.

The next layer, from a depth of about 30 to 150 cm, should also contain noncapillary pores, but a much smaller percentage than the topsoil. These larger pores help to remove excess water but, because of their small numbers, will do it slowly, allowing the rainwater that percolates from above to thoroughly distribute itself in the smaller pores of the entire layer. It is particularly important for this second layer to be well supplied with pores of capillary size. These serve in the movement of water in all directions, especially toward areas where roots are using water.

Soil below a depth of 150 cm affects the growth of most crop plants mainly as a storage of water. If the soil above is well supplied with capillary pores of the proper size, it can lift much water up to the roots. It is therefore desirable to have this lower layer practically free of large pores so that water will drain away only slowly. In case of high groundwater or a very humid climate, a higher percentage of large pores would be needed in all soil layers.

TEMPERATURE

Temperature requirements of crops vary. Diurnal soil-temperature fluctuations are believed to be beneficial to plant growth directly. Fluctuations are also desirable because of vapor pressure differences that will help to place the water where it is needed. Extreme temperature fluctuations are undesirable: in summer because of drying out of the soil, in winter because of water accumulation at the surface due to freezing. Both the rate and the frequency of freezing of the topsoil should be minimized to protect soil structure. One of the problems of soil-temperature management is to maintain temperature levels that are well coordinated with air temperatures. Probably one of the most important aims in the temperate zone is to assist the soil in heating up in the spring so that crops will have a longer growth period.

COLOR

Under certain conditions it may be desirable to change the color of the soil surface in order to modify the absorption and reflection of heat radiation, but generally it is not practical to do so. The effort and expense are out of proportion to the effects upon temperature and moisture obtained. At any rate soil-surface color can exert an effect only while the ground is not covered by vegetation.

RECOGNIZING SOIL PHYSICAL CONDITIONS

Before a definite land use and crop can be assigned to a given piece of land, its physical conditions have to be recognized. This involves mostly those characteristics of the soil that are essentially permanent, including texture and organic matter content of the various horizons of the soil profile, the slope of the land, the depth to groundwater, the possible exposure to flooding, and the specific climatic conditions of the locality. After these facts have been determined the more changeable conditions need to be considered so that all the measures can be taken that will help the production and permanence of the soil.

Both quantitative and field-observation methods exist for this purpose. It is also possible to draw conclusions concerning soil conditions from the species and the appearance of plants growing on the land. It is not the purpose here to discuss these methods, only to point out that any soil treatment should be preceded by a clear recognition of the situation that needs adjustment.

METHODS OF ALTERING SOIL PHYSICAL CONDITIONS

A large number of methods exist for managing moisture, aeration, temperature, and structure of the soil. It has to be recognized that any individual method affects several soil conditions. Irrigation can be used as an example. While its main purpose is to increase the water supply in the soil, it modifies the aeration, temperature, and the color of the soil and may affect its structure, to say nothing of chemical and biologic changes.

For this reason it is impossible to separate the discussion of the methods of modifying soil physical conditions according to individual factors being altered.

LAND USE

The type of crop grown has a profound effect on the soil. Differences in soil cover, root system, tillage and harvesting methods, and the residue left account for this. This is one of the reasons that under certain conditions crop rotations are used in preference to monoculture. Although the cash crop may damage the physical conditions of the soil, the meadow crop may help to improve it. This has been found to be the case for corn on fine-textured soils, especially where stover yields are low or where the corn is harvested for silage.

It is likely, however, that with generous fertilization, proper tillage, and residue management such unfavorable effects of cropping with corn or any other row crop are greatly reduced.

Plant stand and vigor affect temperature and moisture. A dense, luxurious stand requires more moisture for transpiration than a puny stand; it also shades the ground more and keeps the soil cooler. This in turn may reduce evaporation and even cause condensation of atmospheric moisture in the soil.

Land use that keeps the soil permanently covered with vegetation and never tilled will, of course, affect the soil differently than a situation that exposes the soil to rainfall impact and stirs it up mechanically. Forest is probably the land use that leaves the soil in the best condition, whereas pasture soil has a rather compact surface layer as a result of trampling by animals. Obviously the specific conditions of grazing density and climate have a bearing on the influence of pasturing on soil conditions.

HARVESTING METHODS

Generally speaking, the more of the crop is harvested, the more unfavorable it is upon the structure of the soil, as organic matter and plant nutrients are largely removed. Also the equipment used is important.

Modern harvesting machines, such as the combine and the picker-sheller, that only remove the grain and leave straw or stover and cob evenly spread in the field are favorable for physical soil conditions. Unfortunately some of these machines are so heavy that they greatly compact the soil, especially if harvesting is done when the soil is wet.

Pasturing is a special type of harvesting. Whether pasturing is good or bad for the soil depends on the intensity of grazing, the closeness of cropping, and whether the pasturing is done when the soil is wet. Sheep and especially goats are much harder on a pasture than horses and cattle. Sheep and goats eat off the plants directly at ground level, thus reducing the opportunity for photosynthesis and leaving the soil bare and exposed to the impact of rain and wind. If the animals are removed from the pasture when forage growth becomes limited and when the soil is wet, then grazing can be beneficial to the soil.

CROP-RESIDUE MANAGEMENT AND MULCHING

The amount and the management of crop residues are of major importance. A generous amount of residues placed in and above the soil can improve moisture, aeration, structure, and temperature conditions. The methods generally used are green manuring, turning under crop residues with a plow, mulching, application of manure. Time and depth of application and state of decomposition of crop residues are of importance. One method of residue management is burning. This may be justified in subarctic climates where an excessive amount of carbonaceous residues would tie up much of the mineral nitrogen in the soil or in a situation where the use of fertilizer is economically impossible. However, burning is usually detrimental for the physical conditions of the soil.

Mulching, the covering of the soil surface with crop residues or other material such as paper or plastic, has a profound effect on the soil. It protects the soil from rainfall impact, decreases evaporation, and reduces the temperature fluctuations.

Indirectly mulching increases the moisture content of the soil, raises its heat capacity, and frequently results in slower warming up of the soil in the spring. What the effects of a given mulch will be depends on its color and whether it is pervious or impervious to water.

TILLAGE

Tillage affects all physical soil conditions. Historically tillage was necessary to eliminate weed and crop competition and to loosen tight soil so that seed could be planted. Essentially this is still the same today, but power equipment and herbicides have greatly changed the situation.

As we learn more about the requirements of plants, we realize that thorough loosening of the soil is not always desirable. Both growth-chamber research and field experiments have shown that many important crops grow well in fairly dense soil. It is well possible that tillage will be further restricted in the future after we have learned better to cooperate with Nature.

Since weed control is increasingly achieved with herbicides, the main goal of tillage is to bring about a soil structure that is beneficial for plant growth. This includes a firm seedbed in the planting rows with good capillary contact to the subsoil but without surface crust, and a fairly loose structure in the soil between the rows. This allows for ample infiltration and aeration, yet prevents excessive loss of water by evaporation.

The texture and structure of a given soil will determine how this can best be achieved. Medium- to coarse-textured soils with a fair humus content retain a favorable structure for considerable periods and therefore require a minimum of tillage. Soils high in clay and silt have a tendency to "run together" and to restrict the passage of air and water. This is especially true if the soils are acid and low in humus. Such soils require more frequent and more elaborate tillage. In the case of organic soils the danger exists that tillage will result in excessive loosening, which might cause rapid oxidation of the soil substance, poor contact with the subsoil, and losses by wind erosion.

The moldboard plow is probably the most efficient implement to loosen a large volume of soil and to break up a massive soil into smaller units. The outstanding feature of the moldboard plow is the twisting and turning of the soil. If soil is plowed at the optimum moisture condition—moist, but not wet—the soil is broken up along its natural cleavage planes and consequently the resulting aggregates are more stable than where the aggregates are sliced open with a knifelike implement. Nevertheless the loose structure resulting from plowing does not last long.

The plow also covers all plants and plant residues and leaves a clean field, simplifying subsequent tillage and planting operations. The disadvantages of the plow are the lack of protection of the bare soil and the troweling action of the plowshare—especially if it is dull—which compacts the soil below it, creating the notorious plow pan (Fig. 10-1). For this reason, it is well to vary the depth of plowing from time to time. When the subsoil is in friable condition, plowing can be done relatively

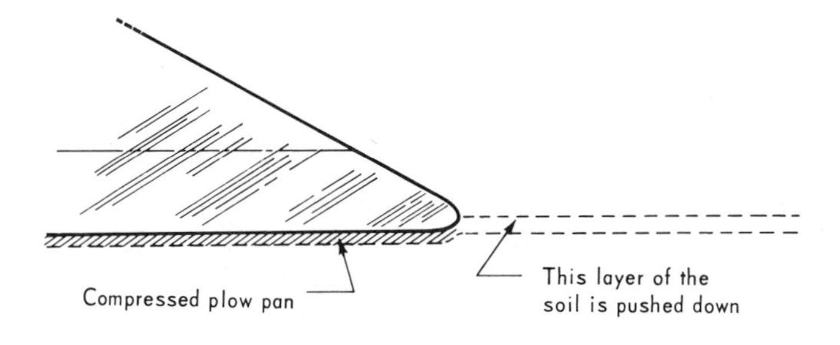

Fig. 10-1 A dull plowshare compresses the soil.

deeply. If it is plastic, a shallow depth should be chosen in order to avoid puddling the subsoil. Under many conditions planting can follow plowing immediately or after smoothing out the soil surface with a simple tool that can be pulled behind the plow. Such "minimum tillage" retains the loose structure created by the moldboard plow, reduces the weed problem, and costs less than the more elaborate tillage systems. The change from "conventional tillage" to "minimum tillage" has been the consequence of the increased power of farm tractors.

The disk serves to cut clods and plant residues and creates a rather satisfactory seedbed. Most disks do not penetrate the soil very deeply. They have the tendency to compact the soil underneath. Probably the greatest defect of the disk is that it slices the soil and exposes the unprotected surfaces of the aggregates to the influence of the weather. This is the reason that soil that is disked has much more tendency to break up and to become dense than plowed land.

Cultivators open up the soil without deleterious effects upon the lower strata. They are not as efficient in loosening the ground as the plow and they also turn under neither the surface soil nor the crop residues. (This may be an advantage or disadvantage.) The smaller cultivators can be used only on land that has been tilled previously and where there are not too many crop residues. The larger ones are well adapted to open up massive soil and do not clog up with residues because the shares have high clearance and are put far enough apart. These large cultivators, especially when combined with a fertilizer distributor system, serve well for mulch tillage. Field cultivators do not leave all the residues on the surface but usually enough to give the soil a satisfactory protection. Field cultivators are best adapted to prepare ground with little or no living vegetation, but not with sod, for the following crop. Probably the greatest objection to the exclusive use of field cultivators as the main tillage tool—eliminating the moldboard plow—is that it is difficult, if not impossible, to place fertilizer deep enough so that it is in moist soil throughout the growing season. Because of this it is advisable to incorporate extra amounts of potassium and phosphate with a plow in one year and to use the field cultivator in the intervening years. The higher the plant nutrient level of the soil, the longer can be the interval between the times fertilizer is plowed under.

Subsoilers are used to open tight, impervious subsoils. Their effects are frequently unsatisfactory because the clayey subsoil runs together after it becomes thoroughly wetted. It has been tried to keep the subsoil slits open by placing fertilizer in them, thus attracting root growth, but with only limited success. Filling these subsoil slits with crop residues—the so-called vertical mulching—has been more effective (Fig. 10-2).

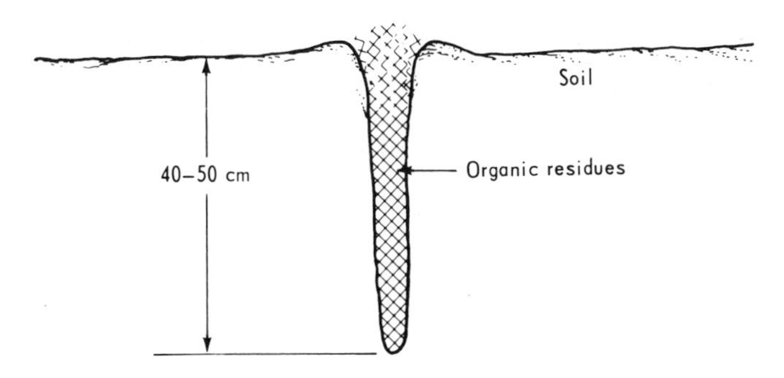

Fig. 10-2 Vertical mulch channel, schematic cross section: For quick entry of water the organic residues in the channel must reach to the surface and must not be covered with soil. The vertical mulch channels are placed about 5 m apart.

Land levelers are generally used only where flood irrigation is planned. However, they can serve well to smooth out slopes that have become irregular through erosion. In this way surface water will be better distributed and will not so easily have the tendency to concentrate and cause erosion.

Special attention should be given the weight of tractors and any other equipment passing over the ground, or maybe the ratio between the weight and area of support. The greater the weight per unit area, the greater will be the damage done to soil structure. However, also the area of support as such is important. Where the weight per unit area is the same, the larger area of support as such is important, since it will be more damaging than the smaller one because the sides will help to support the compressed soil (arch action). Since tractors must have traction as well as power, the problems of compaction cannot be solved solely by increasing the ratio of tractor power to tractor weight. Low pressure tires help to reduce the weight per unit area and also to increase traction. The difficulty is to operate such tires on hard-surfaced roads.

DRAINAGE

Soils are drained for two purposes: to increase the rate of air renewal and to leach salts from irrigated soils. In both cases water has to be removed from soils.

While superficial runoff assists in removing excess water, the word "drainage" generally refers to the removal of water that has percolated through the soil. Both open ditches and tiles are used for this purpose. They serve to lower the water table and put the soil water under tension. Many tiles are placed at a depth of around 100 cm. Consequently, after the water table has dropped down to this depth, the moisture near the surface will be under a tension of 100 cm ($1/10$ bar), while at a depth of 50 cm it will be at a tension of 50 cm. This is equal to the *aeration-porosity limit.* All connected pores larger than 0.06 mm in diameter will be free of water. This assumes that water has not been lost by evaporation or by transpiration. It also assumes that the soil is near the tile or ditch. If the tiles are a fair distance apart, as they normally are, the groundwater level in the middle between the tiles will be nearer the surface (Fig. 10-3). The less pervious the soil, the steeper will be the curvature of the groundwater level and the shallower will it be between the tiles. Consequently tiles are spaced closer together on fine-textured soils than on coarse-textured soils.

If the lack of air in the soil is only the result of a high general groundwater table, removing of the excess water by drainage will generally rectify the situation. As the water table is lowered to a desired depth, air enters the soil and the growing conditions for plants are greatly improved. Frequently the lack of air, however, results from an accumulation of water in a soil that is so tight that water cannot percolate fast enough. Also in this case, drainage by tile and ditch can be helpful, but the slow movement of water through such a soil makes the task more difficult.

Since tiling or ditching cannot even remove water to "field capacity," fine-textured soils cannot be overdrained. However, sands have such a small amount of available-water-holding capacity that drainage has to be done with caution. Also mucks can be overdrained, especially because they oxidize too quickly, if they are drained to a depth of 100 cm. This results in loss of substance and in subsidence of the surface of the land. Therefore drain tiles in muck land are usually placed shallower than 100 cm.

In areas where the water is saline, leaching of salts from irrigated soils is necessary to maintain the productivity. In this case removal of

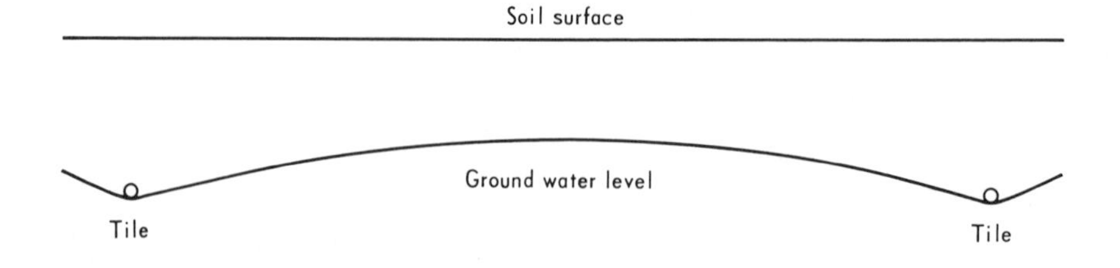

Fig. 10-3 The groundwater level in a tiled field: The rate of drop of the groundwater level depends on the permeability of the soil and on the distance the water has to travel to the tile.

the leaching water is the main purpose of drainage. Therefore the tiles have to be placed much deeper—200 to 250 cm—than where the purpose is only improvement of aeration.

A monograph published by the American Society of Agronomy (Luthin, ed., 1957) discusses all important phases of soil drainage.

IRRIGATION

Irrigation—the application of water to the soil—should be considered whenever natural soil moisture is limiting plant growth and crop production. Not in all cases would it be wise to use irrigation. Since irrigation is always costly, it is economical only if it can be expected that the returns will exceed the investment and the operating costs.

The application of water to the soil can be handled in a variety of ways. The three basic methods are surface irrigation, sprinkler irrigation, and subirrigation.

In surface irrigation the water flows by gravity over the field. In its simplest form runoff from a permanent or temporary stream is diverted by a series of low dams across the valley so that the water can cover a large area. Thus the water will infiltrate instead of being lost to the farm. A slightly more elaborate method is the so-called border dyke irrigation. Parallel dykes are used to allow water to flow down a slight slope and to cover the entire area between the dykes. The land surface must be smooth and the slope uniform so that all of the land gets covered by water and no pockets develop. This system is adapted to pasture and close-growing crops. A method that allows more uniform distribution of the water is furrow irrigation. Water is made to flow along every furrow or every second furrow so that the ridges remain free. This has the advantage that the structure and aeration of the soil in the ridges remain unimpaired.

Gravity is the motive force in the case of surface irrigation, and the transportation of the water above ground is in canals and ditches. For sprinkler irrigation a system of pipes is needed and sufficient pressure to distribute the water through these pipes and from the sprinkler nozzles to the plants. This makes sprinkler irrigation more costly per unit of water. It is, therefore, used chiefly where only a small fraction of the water needs of a crop has to be satisfied by irrigation.

Subirrigation is a very practical and relatively inexpensive method of applying water to soil. It is sometimes called *controlled drainage* because the same tile system or drainage ditch that is used for removal of excess water serves to bring water to the field during drought periods. Subirrigation is adapted only to nearly flat areas with highly pervious subsoils, such as sands and muck, and a high groundwater table or an impervious substratum at greater depth. During the period the soils are too wet, the drain tiles and drainage ditches function in the ordinary manner. Once the water level begins to fall down too far to supply the plant roots by capillary action, water is allowed to flow into the tile outlets or drainage ditches and is distributed throughout the field.

Several factors are important in determining whether irrigation should be used and what type of irrigation will fit best. Whenever there is a lack of moisture in the soil during the period when the temperature is favorable for crop growth, irrigation should be considered. This situation depends on a combination of the water needs of the crop (C), the storage of water at planting time (S), the precipitation during the growing season (P), and the losses of water by evaporation from the soil (E), the transpiration by weeds (W), runoff (R), and drainage (D). If during the growing season $S + P - E - W - R - D > C$, irrigation is not needed. If it is smaller than C for part or all of the period, the technical and economic feasibility of irrigation should be investigated. The storage of water in the soil at planting time depends both on the climate and on the physical nature of the soil itself. Where ample precipitation precedes the planting time, it is likely that the soil is moist to a considerable depth. The amount of water in the soil depends on its available-water-holding capacity. This varies from 5 to 25 percent of the volume and is generally around 15 percent. This means that a soil that has reached field capacity to a depth of 1 m would contain 15 cm of available water. This is about one-third of the water needed for a good corn crop. How deep roots will actually penetrate the soil depends both on the moisture and on the oxygen in the soil.

Long-term precipitation records can tell whether or not the water stored in the soil during the growing period will remain at a tension low enough to permit vigorous crop growth. Where only occasional periods of inadequate water supply occur, it is usually not economical to use irrigation except for special high-value crops. If summer droughts are a regular occurrence and only small amounts of water make the difference between a good crop and a failure, the use of sprinklers is probably advisable for such supplemental irrigation. While it is fairly expensive per unit of water applied, the land does not have to be graded and dyked and few, if any, ditches are needed. The sprinkler equipment can be hauled from place to place.

In an area of permanent water deficiency, surface irrigation is best adapted. Regular irrigations are needed and every field has to be graded and ditches have to be placed at the upper end of each of them so that water can readily be brought to them. The best use of the water—and of the labor—is made if the maximum amount is applied each time that is compatible with the purpose. Since soil soaks up water to field capacity before any of it can penetrate further down, it is best to regulate the application so that the soil gets wet to the depth to which root growth is expected.

The type of the soil and the quantity and quality of the available water have a great influence in determining whether irrigation will be a success or a failure. In an area where most of the water needed by the crops comes from rain, good irrigation water is usually available in sufficient amounts. Under such conditions soil structure is the main item that determines the feasibility of irrigation. In such a climate aeration of the soil is frequently just as great a problem as drought. Unless the soil has a large aeration capacity, the possibility exists that the plants in an irrigated soil will suffer from lack of oxygen, especially if a rainy period follows an irrigation. The greater the available-water-holding capacity of a soil, the less important becomes supplemental irrigation. This is true for the finer-textured soils, whereas sands and other coarse-textured soils have usually both low available-water-holding capacity and high aeration capacity. Figure 10-4 gives typical cases of soils of both kinds.

In arid or semiarid climates both water and soil can pose great obstacles to successful irrigation. Assuming the quantity of available water is sufficient, the small amount of rainfall in the area may cause the water to be charged with soluble salts, especially sodium salts. If such water is used for irrigation, the salts will accumulate in the soil while the water itself is used for transpiration and evaporation. In some cases soils become unproductive and practically sterile after a few years of irrigation. Water with a high percent of sodium disperses the clay and the soil becomes nearly impervious to both air and water. The high reaction caused by sodium carbonate makes many of the plant nutrient elements unavailable and dissolves the organic matter to bring about the dreaded condition known as *black alkali.* Even a less pronounced saline or alkaline situation can stunt or kill plants. Whenever saline water is used for irrigation, it is necessary to provide for drainage. Once during the year—preferably in late winter when the water supply is most abundant and its salt content the lowest—heavy irrigation must be applied in order to leach out the soluble salts that may have accumulated. For this reason it is best to place the drain tiles quite deep—2 to 2.5 m—so that sufficient tension exists in the root zone to remove much of the applied water. If sodium causes dispersion of the soil, calcium sulfate is applied before the irrigation. The calcium replaces the sodium and helps in the flocculation of the clay. If for any reason drainage is impossible, it is best not to set up surface irrigation, because the hazard of salinity exists. Accumulation of salts in the soil can make the production of crops impossible.

Since climate, soil, and crops vary from one irrigation project to

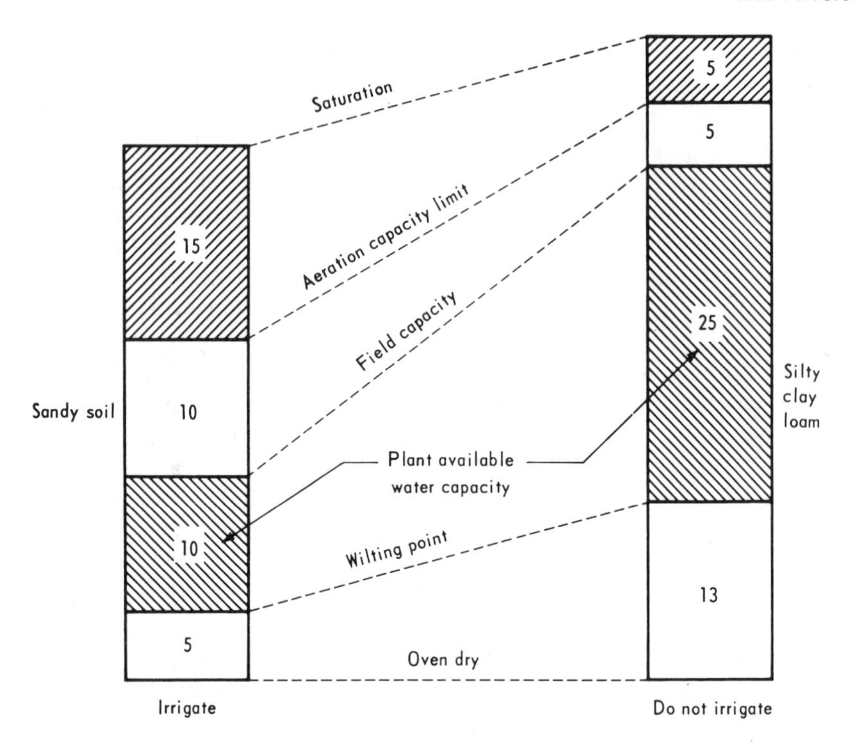

Fig. 10-4 Soil texture and structure determine advisability of irrigation in a humid or semihumid climate: Figures denote percent by volume. Adaptation of soils to supplemental irrigation depends on pore space distribution. The small volume of plant-available-water-holding capacity in the sandy soil indicates its droughtiness and calls for irrigation. Fifteen percent of large pore space guarantees adequate aeration. The 25 percent of plant-available-water-holding capacity of the silty clay loam can furnish the plants ample moisture, even in a dry spell. The small amount of large pores points to aeration problems, if irrigation is used.

the next, no general rules can be given as to the frequency and the amount of water to be applied. However, it is always advisable to irrigate while the plants still find enough water for best growth. Because at the same tension level fine-textured soils contain more water than coarse-textured soils, irrigation should start in the latter at a lower tension. Figure 10-5 illustrates this graphically. The irrigation point varies from ½ to 5 or 7 atmospheres. The problems of irrigation of agricultural lands are discussed extensively in a monograph published by the American Society of Agronomy (Hagan *et al.*, eds., 1967).

CHEMICAL TREATMENTS

Of the various chemicals that are used on the land, only the *soil conditioners* are used with the specific purpose of improving the physical

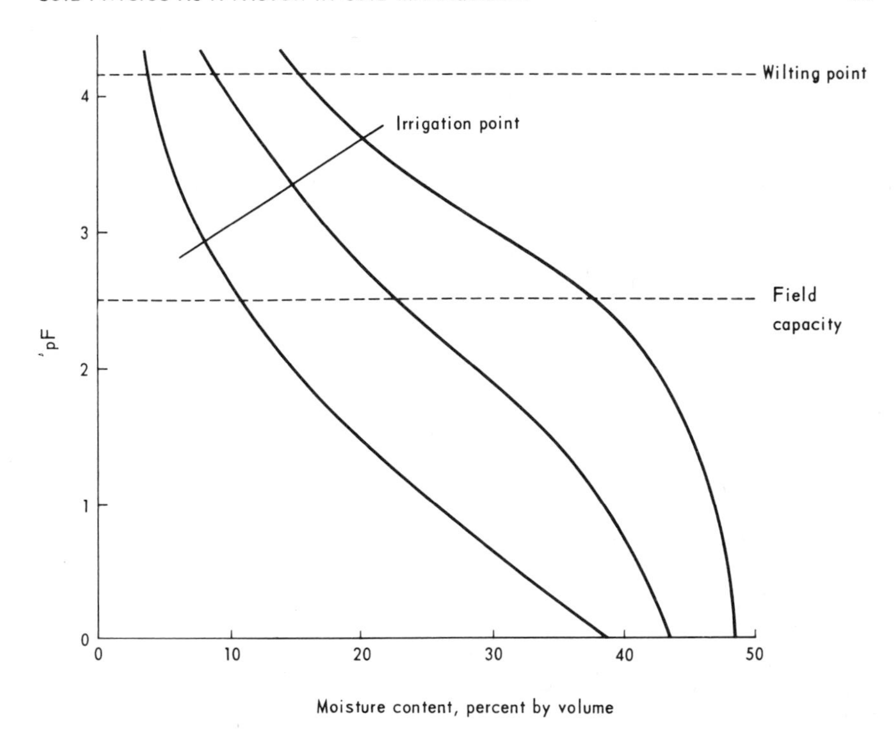

Fig. 10-5 The amount of available water left in the soil determines when to irrigate: The irrigation point is not equipotential for all soils. It crosses the desorption curves of the various soils at a location, where between 4 and 6 percent by volume of available water is left.

conditions of the soil. Most of them consist of large organic molecules that serve to tie many clay particles together. Technically efficient, they have found only little application in agriculture due to their high cost. It is well possible that a more economic fabrication and development of specifically adapted methods of use will increase their popularity.

Fertilizers and lime have the obvious advantage of increasing plant growth and thus supplying more residues and more energy for microbes. An improved and more stable soil structure is the result.

Herbicides have the indirect effect on soil physical conditions that the reduced weed growth requires less tillage. The heavy equipment that is used to apply herbicides as well as insecticides may damage soil structure, especially if it is operated on wet ground. The soils in the alleys of orchards where a regular spray program is followed are sometimes exceedingly dense. Herbicides and insecticides, especially their prolonged use, may have a deleterious effect upon the microbial activity in the soil and thus damage soil structure indirectly.

REFERENCES

Hagan, R. M., H. R. Haise, and T. W. Edminster (eds.): "Irrigation of Agricultural Lands," no. 11 in the Series Agronomy, American Society of Agronomy, Inc., 1967.

Luthin, J. N. (ed.): "Drainage of Agricultural Lands," no. 7 in the Series Agronomy, Am. Soc. of Agron., Inc., 1957.

Spain, J. M., and D. L. McCune: Something New in Subsoiling, *Agron. J.*, vol. 48, pp. 192–193, 1956.

INDEX

INDEX